Irrigation Engineering

Irrigation Engineering

Lucias Chapman

R CALLISTO REFERENCE

www.callistoreference.com

Callisto Reference,
118-35 Queens Blvd., Suite 400,
Forest Hills, NY 11375, USA

Visit us on the World Wide Web at:
www.callistoreference.com

ISBN: 978-1-64116-538-9 (Hardback)

Cataloging-in-Publication Data

Irrigation engineering / Lucias Chapman.
        p. cm.
Includes bibliographical references and index.
ISBN 978-1-64116-538-9
1. Irrigation engineering. 2. Hydraulic engineering. 3. Agricultural engineering. I. Chapman, Lucias.
TC805 .I77 2022
627.52--dc23

Table of Contents

	Preface	**VII**
Chapter 1	**Introduction to Irrigation**	1
	• Irrigation System	4
Chapter 2	**Fundamental Aspects of Irrigation Engineering**	14
	• Planning of an Irrigation Project	14
	• Financial Analysis of Irrigation Project	16
	• Types of Canal Construction Machineries	23
	• Assessment of Charges for Irrigation Water	29
	• Irrigation Efficiency	30
	• Design, Construction and Maintenance of Irrigation Reservoirs	36
	• Choice and Selection of Irrigation Methods	48
	• Crop Water Requirement in Irrigation	51
	• Irrigation Scheduling	55
	• Construction of an Evaporation Pan for Irrigation Scheduling	57
Chapter 3	**Irrigation: Operations, Processes and Systems**	61
	• Water Conveyance and Distribution Canals	61
	• Open Channel Flow	70
	• Design of Open Channel	76
	• Irrigation Scheduling for Regulated Deficit Irrigation	81
	• Irrigation System Controllers	85
	• Automatic Plant Irrigation System	93
	• Automated Rain Barrel Drip System	96
	• Drip Irrigation System Plan	99
	• Evaluation of Rotating Head Sprinklers and Operation of Sprinkler System	100
	• Design and Operation of Underground Pipeline System	107
Chapter 4	**Land Grading and Survey in Irrigation**	112
	• Land Evaluation	112
	• Surveying for Construction of Irrigation Projects	115
	• Land Grading for Irrigation	120
	• Land Levelling	123
	• Contour Bench Levelling and Earthwork Quantities Computation	138

Chapter 5   **Irrigation Canals**   144

- Canal Lining   146
- Design of Irrigation Canals   151
- Minimum Cost Design of Lined Canal Sections   164
- Basic Components of Barrage   172

Chapter 6   **Irrigation Hydraulics and Fluid Mechanics**   177

- Irrigation Hydraulics   177
- Irrigation Water Measurement   195
- Volumetric Measurement of Irrigation Water use   199
- Water Lifting for Irrigation   202
- Hydraulics of Drip Irrigation System   224
- Water Filtration for Irrigation Systems   230
- Surface Irrigation Hydraulics   232

Chapter 7   **Irrigation System Design**   239

- Micro Irrigation Systems Design Considerations   241
- Border-check Irrigation Design   247
- Design of Sprinkler Irrigation System   251
- Design of Drip System   259

Permissions

Index

Preface

The application of limited amounts of water to plants at required intervals is known as irrigation. It helps in the growth of crops and in maintaining landscapes. It also plays a vital role in re-vegetating the disturbed soils in dry areas. The branch of engineering which deals with harnessing and controlling the water which is obtained from natural sources and distributing it for agricultural purposes is known as irrigation engineering. There are various types of irrigation methods such as micro-irrigation, drip irrigation, and sprinkler or overhead irrigation. Micro-irrigation uses a piped network to distribute water under low pressure. Drip irrigation is a system that drops water directly at the plant's roots. Sprinkler or overhead irrigation is a system where water is piped to the central locations within the field and distributed by high-pressure overhead sprinklers. This book provides comprehensive insights into the field of irrigation engineering. It will serve as a reference to a broad spectrum of readers. Those in search of information to further their knowledge will be greatly assisted by this textbook.

A foreword of all chapters of the book is provided below:

Chapter 1 - Applying a specific amount of water to plants when needed is known as irrigation. It helps in the growth of agricultural crops and in maintaining landscapes. An irrigation system is made up of several components such as an intake structure, conveyance system, field application system and drainage system. This is an introductory chapter which will introduce briefly all these significant aspects of irrigation and irrigation systems.; **Chapter 2** - The branch of engineering which focuses on the analysis and design of irrigation systems such as barrages, canals, drains and dams is known as irrigation engineering. Some of the important aspects of this field are planning irrigation projects, assessing charges for irrigation water and selecting irrigation methods. The chapter closely examines these key aspects of irrigation engineering to provide an extensive understanding of the subject.; **Chapter 3** - There are various processes which are a part of irrigation such as water conveyance, irrigation scheduling and designing open channels. Some of the systems which are used in irrigation are automatic plant irrigation system, automated rain barrel drip system and underground pipeline system. This chapter discusses in detail these processes and systems related to irrigation.; **Chapter 4** - Reshaping the land surface to planned grades for the purpose of irrigation and subsequent drainage is known as land grading. Leveling the land is an important aspect of irrigation since it ensures that the depths and discharge variations over the field are relatively uniform. The diverse applications of land grading, surveying and leveling have been thoroughly discussed in this chapter.; **Chapter 5** - An irrigation canal is a man-made waterway to transport water for agricultural purposes. The impermeable layer which is provided for the bed and sides of canal for the purpose of improving the life and discharge capacity of canals is known as canal lining. The chapter closely examines the key concepts related to canals such as their designing and lining.; **Chapter 6** - Irrigation hydraulics deals with the designing and maintenance of irrigation systems in an economical and efficient manner using the principles from the field of hydraulics. The topics elaborated in this chapter will help in gaining a better perspective about the branches of irrigation hydraulics such as irrigation water management and surface irrigation hydraulics.; **Chapter 7** - The design of an irrigation system is aimed at minimizing damage and losses to soil, water, plant, air and animal resources. It also seeks to reduce deep drainage and runoff. Border check irrigation design and micro irrigation system design are a few examples of different irrigation systems designs. This chapter discusses in detail these theories and methodologies related to the design of irrigation systems.

At the end, I would like to thank all the people associated with this book devoting their precious time and providing their valuable contributions to this book. I would also like to express my gratitude to my fellow colleagues who encouraged me throughout the process.

Lucias Chapman

Chapter 1

Introduction to Irrigation

Applying a specific amount of water to plants when needed is known as irrigation. It helps in the growth of agricultural crops and in maintaining landscapes. An irrigation system is made up of several components such as an intake structure, conveyance system, field application system and drainage system. This is an introductory chapter which will introduce briefly all these significant aspects of irrigation and irrigation systems.

Irrigation is the artificial application of water to land for the purpose of agricultural production. Effective irrigation will influence the entire growth process from seedbed preparation, germination, root growth, nutrient utilisation, plant growth and regrowth, yield and quality.

The key to maximising irrigation efforts is uniformity. The producer has a lot of control over how much water to supply and when to apply it but the irrigation system determines uniformity. Deciding which irrigation systems is best for your operation requires a knowledge of equipment, system design, plant species, growth stage, root structure, soil composition, and land formation. Irrigation systems should encourage plant growth while minimising salt imbalances, leaf burns, soil erosion, and water loss. Losses of water will occur due to evaporation, wind drift, run-off and water (and nutrients) sinking deep below the root zone.

Proper irrigation management takes careful consideration and vigilant observation.

Value of Irrigation

Irrigation allows primary producers:

- To grow more pastures and crops.

- To have more flexibility in their systems/operations as the ability to access water at times when it would otherwise be hard to achieve good plant growth (due to a deficit in soil moisture) is imperative. Producers can then achieve higher yields and meet market/seasonal demands especially if rainfall events do no occur.

- To produce higher quality crops/pastures as water stress can dramatically impact on the quality of farm produce.

- To lengthen the growing season (or in starting the season at an earlier time).

- To have 'insurance' against seasonal variability and drought.

- To stock more animals per hectare and practice tighter grazing management due to the reliability of pasture supply throughout the season.

- To maximise benefits of fertiliser applications. Fertilisers need to be 'watered into' the ground in order to best facilitate plant growth.

- To use areas that would otherwise be 'less productive'. Irrigation can allow farmers to open up areas of their farms where it would otherwise be 'too dry' to grow pasture/crops. This also gives them the capability to carry more stock or to conserve more feed.

- To take advantage of market incentives for unseasonal production.

- To have less reliance on supplementary feeding (grain, hay) in grazing operations due to the more consistent supply & quality of pastures grown under irrigation.

- To improve the capital value of their property. Since irrigated land can potentially support higher crops, pasture and animal production, it is considered more valuable. The value of the property is also related to the water licensing agreements or 'water right'.

- To cost save/obtain greater returns. The cost benefits from the more effective use of fertilisers and greater financial benefits as a result of more effective agricultural productivity (both quality and quantity) and for 'out of season' production are likely.

Choosing an Irrigation System

There is a huge diversity in the types of irrigation technologies/systems used, which is attributable to,

- Variations in soil types.

- Varying topography of the land.

- Availability of power sources.

- Availability of water.

- Sources of water.

- The period of time when the system was installed.

- The size of the area being irrigated.

- On farm water storage capacity.

- Availability of labour/financial resources.

Source of Irrigation Water

The vast majority of irrigation water use is pumped directly from a water source (river, creek, channel, drag line, hole, dam or bore).

Irrigation Scheduling

Irrigation scheduling is the process by which an irrigator determines the timing and quantity of water to be applied to the crop/pasture. The challenge is to estimate crop water requirements for different growth stages and climatic conditions.

To avoid over or under watering, it is important to know how much water is available to the plant, and how efficiently the plant can use it. The methods available to measure this include:

- Plant observation,

- Feel and appearance of the soil,

- Using soil moisture monitoring devices; or

- Estimating available water from weather data.

Plant observation:	Visible changes in plant characteristics, such as leaf colour, curling of the leaves and ultimately wilting can be useful guides to indicate plant moisture stress, and hence the need for irrigation. Productivity may be lowered, particularly if moisture depletion is allowed to the point where wilting occurs. The moisture status of plants can also be measured using sap flow sensors (used mainly for research), infra-red guns (used in the cotton industry) and pressure bombs (which measure leaf water potential).
Feel and appearance of the soil:	Visual observation and feel of the soil is used to monitor moisture levels of paddocks and hence their ability to sustain plant growth. A soil sample can be obtained by using a soil probe, auger or spade. By squeezing soil into a ball, observing the appearance of the ball and creating a ribbon of soil between the thumb and forefinger, soil moisture can be estimated. For example, At 75 % field capacity, (i) Sands and sandy loams - are slightly coherent, will form a weak ball under pressure but breaks easily, (ii) Loams, clay loams and clays- are coherent, soil has a slick feeling and ribbons easily, and will not roll into long thin rods 2.5 diameter, and (iii) Comment - there is adequate water and air for good plant growth At 0-25 % field capacity (or wilting point), (i) Sands and sandy loams - are dry, loose, flows through fingers, (ii) Loams, clay loams and clays are crumbly and powdery, small lumps break into powder, and will not ball under pressure, and (iii) Comment - plants desperately need watering and will die soon.
Weather based data:	There are two weather - based scheduling systems used to measure the amount of water lost from a crop. These are: (i) Evaporation from an open water surface -gives some indication of crop water use (the latter is generally lower), or (ii) Historical climate data such as relative humidity, temperature, wind speed and sunshine hours.
Soil moisture monitoring:	Soil moisture can be measured as a suction or volume of water. This idea is applicable to how much force a plant can exert on the soil to extract the amount of water it needs for growth. Soil moisture suction can be used as a measure of plant stress and for that reason it is a handy tool for growers to use in scheduling their irrigations. Soil moisture monitoring devices include tensiometers and resistance (measure soil moisture suction) or neutron probe, EnviroScan, Gopher, Time Domain Reflectometry (TDR), DRW Microlink and Aquaflex (measure soil moisture content).

Problems

- While irrigation has provided a number of important benefits the potential drawbacks of over/under watering include.

- Under-watering.

- Loss in market value through yield reduction.

- Reduction in fruit size and quality.

Over-watering

- Unwanted vegetative growth.

- Losses of valuable water to the watertable.

- Irrigation water travelling over soil can cause erosion. The excessive displacement of the top soil can also affect soil fertility (and hence crop yields), it may also clog drainage ditches and streams (silting), harm aquatic habitats, foul waters used for recreational activities, and increases the need for water treatments.

- Irrigation can cause pesticides, pathogens and weeds to spread during irrigation.

- Cause runoff.

- Increased operational costs (labour, pumping, cost of water).

- Leaching of nutrients (eg. salt, phosphorus) may lead to algal growth, salinity an nitrate build ups (poisoning) elsewhere in the catchment.

- Downgraded product quality and reduced yield.

- Higher operational costs for the producer (hence, reduced profits).

- Pressue on water resources with the Increasing demand for water use by urban dwellers.

Irrigation System

The irrigation system consists of a (main) intake structure or (main) pumping station, a conveyance system, a distribution system, a field application system and a drainage system.

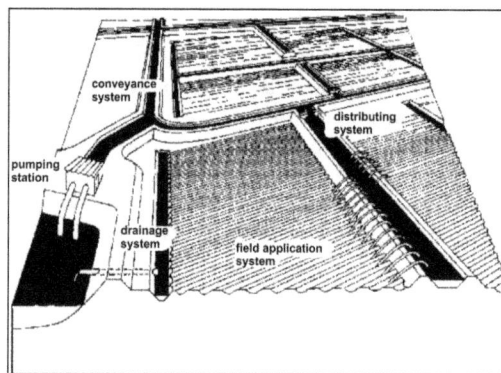

An irrigation system.

- The (main) intake structure, or (main) pumping station, directs water from the source of supply, such as a reservoir or a river, into the irrigation system.

- The conveyance system assures the transport of water from the main intake structure or main pumping station up to the field ditches.

- The distribution system assures the transport of water through field ditches to the irrigated fields.

- The field application system assures the transport of water within the fields.

- The drainage system removes the excess water (caused by rainfall and/or irrigation) from the fields.

Main Intake Structure and Pumping Station

Main Intake Structure

The intake structure is built at the entry to the irrigation system. Its purpose is to direct water from the original source of supply (lake, river, reservoir etc.) into the irrigation system.

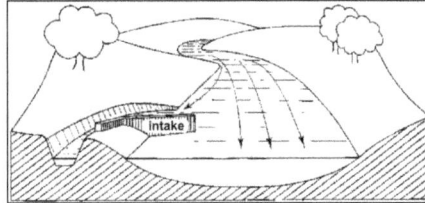
An intake structure.

Pumping Station

In some cases, the irrigation water source lies below the level of the irrigated fields. Then a pump must be used to supply water to the irrigation system.

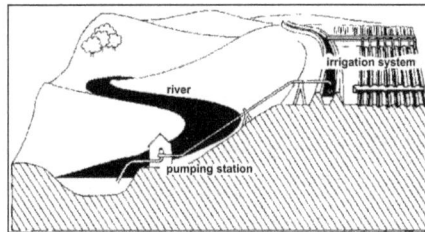
A pumping station.

There are several types of pumps, but the most commonly used in irrigation is the centrifugal pump.

The centrifugal pump consists of a case in which an element, called an impeller, rotates driven by a motor. Water enters the case at the center, through the suction pipe. The water is immediately caught by the rapidly rotating impeller and expelled through the discharge pipe.

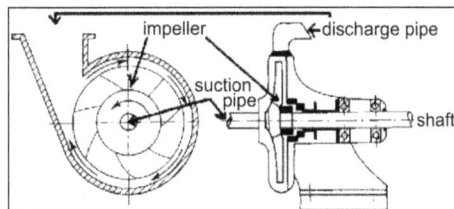
Diagram of a centrifugal pump.

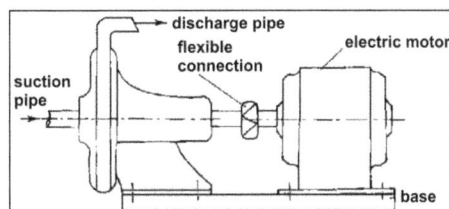
Centrifugal pump and motor.

The centrifugal pump will only operate when the case is completely filled with water.

Conveyance and Distribution System

The conveyance and distribution systems consist of canals transporting the water through the whole irrigation system. Canal structures are required for the control and measurement of the water flow.

Open Canals

An open canal, channel, or ditch, is an open waterway whose purpose is to carry water from one place to another. Channels and canals refer to main waterways supplying water to one or more farms. Field ditches have smaller dimensions and convey water from the farm entrance to the irrigated fields.

Canal Characteristics

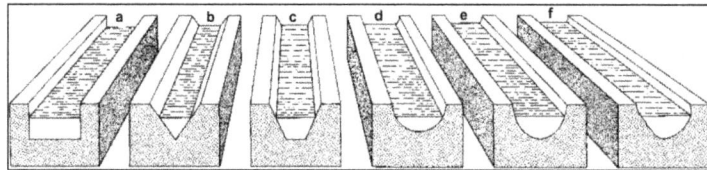

According to the shape of their cross-section, canals are called; (a) rectangular, (b) triangular, (c) trapezoidal, (d) circular, (e) parabolic, (f) irregular or natural.

The most commonly used canal cross-section in irrigation and drainage, is the trapezoidal cross-section. For the purposes of this publication, only this type of canal will be considered.

The typical cross-section of a trapezoidal canal.

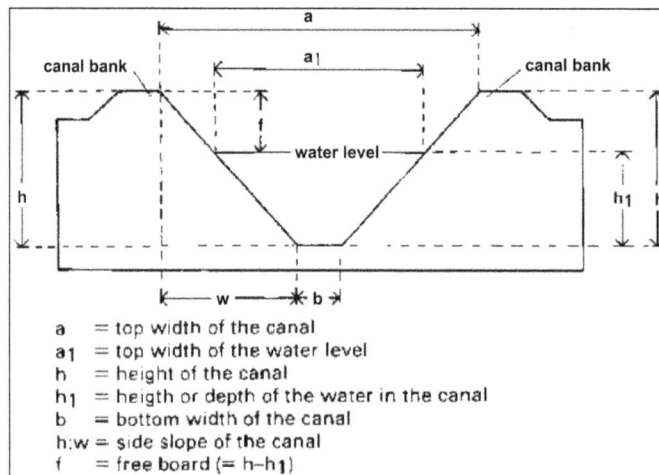

A trapezoidal canal cross-section.

The freeboard of the canal is the height of the bank above the highest water level anticipated. It is required to guard against overtopping by waves or unexpected rises in the water level.

The side slope of the canal is expressed as ratio, namely the vertical distance or height to the horizontal distance or width. For example, if the side slope of the canal has a ratio of 1:2 (one to two), this means that the horizontal distance (w) is two times the vertical distance (h).

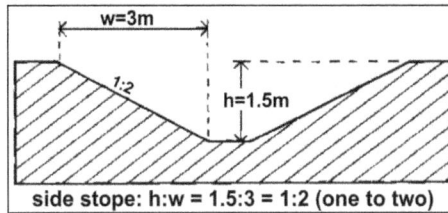

A side slope of 1:2 (one to two).

The bottom slope of the canal does not appear on the drawing of the cross-section but on the longitudinal section. It is commonly expressed in percent or per mil.

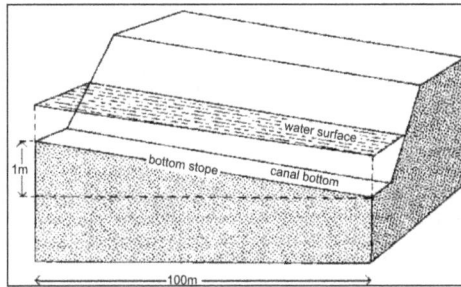

A bottom slope of a canal.

An example of the calculation of the bottom slope of a canal is given below:

$$\text{the bottom slope } (\%) = \frac{\text{height differene}(\text{metres})}{\text{horizontal distance}(\text{metres})} \times 100 = \frac{1\,\text{m}}{100\,\text{m}} \times 100 = 1\,\%$$

Or

$$\text{the bottom slope } (\%) = \frac{\text{height differene}(\text{metres})}{\text{horizontal distance}(\text{metres})} \times 1000 = \frac{1\,\text{m}}{100\,\text{m}} \times 1000 = 10\,\%$$

Earthen Canals

Earthen canals are simply dug in the ground and the bank is made up from the removed earth.

The disadvantages of earthen canals are the risk of the side slopes collapsing and the water loss due to seepage. They also require continuous maintenance in order to control weed growth and to repair damage done by livestock and rodents.

Lined Canals

Earthen canals can be lined with impermeable materials to prevent excessive seepage and growth of weeds.

Lining canals is also an effective way to control canal bottom and bank erosion. The materials mostly used for canal lining are concrete (in precast slabs or cast in place), brick or rock masonry and asphaltic concrete (a mixture of sand, gravel and asphalt).

The construction cost is much higher than for earthen canals. Maintenance is reduced for lined canals, but skilled labour is required.

Canal Structures

The flow of irrigation water in the canals must always be under control. For this purpose, canal structures are required. They help regulate the flow and deliver the correct amount of water to the different branches of the system and onward to the irrigated fields.

There are four main types of structures: erosion control structures, distribution control structures, crossing structures and water measurement structures.

Erosion Control Structures

a. Canal erosion

Canal bottom slope and water velocity are closely related, as the following example will show.

A cardboard sheet is lifted on one side 2 cm from the ground. A small ball is placed at the edge of the lifted side of the sheet. It starts rolling downward, following the slope direction. The sheet edge is now lifted 5 cm from the ground, creating a steeper slope. The same ball placed on the top edge of the sheet rolls downward, but this time much faster. The steeper the slope, the higher the velocity of the ball.

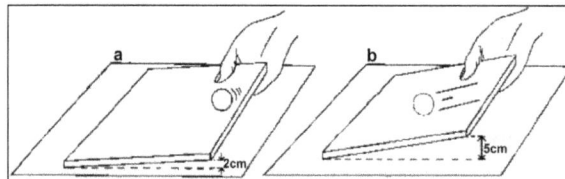
The relationship between slope and velocity.

Water poured on the top edge of the sheet reacts exactly the same as the ball. It flows downward and the steeper the slope, the higher the velocity of the flow.

Water flowing in steep canals can reach very high velocities. Soil particles along the bottom and banks of an earthen canal are then lifted, carried away by the water flow, and deposited downstream where they may block the canal and silt up structures. The canal is said to be under erosion; the banks might eventually collapse.

b. Drop structures and chutes

Drop structures or chutes are required to reduce the bottom slope of canals lying on steeply sloping land in order to avoid high velocity of the flow and risk of erosion. These structures permit the canal to be constructed as a series of relatively flat sections, each at a different elevation.

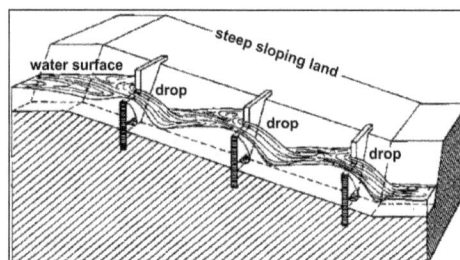
Longitudinal section of a series of drop structures.

Drop structures take the water abruptly from a higher section of the canal to a lower one. In a chute, the water does not drop freely but is carried through a steep, lined canal section. Chutes are used where there are big differences in the elevation of the canal.

Distribution Control Structures

Distribution control structures are required for easy and accurate water distribution within the irrigation system and on the farm.

a. Division boxes

Division boxes are used to divide or direct the flow of water between two or more canals or ditches. Water enters the box through an opening on one side and flows out through openings on the other sides. These openings are equipped with gates.

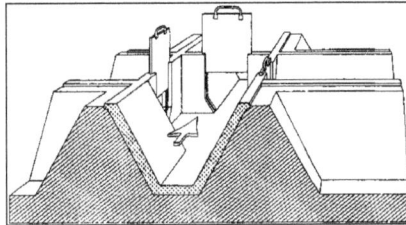

A division box with three gates.

b. Turnouts

Turnouts are constructed in the bank of a canal. They divert part of the water from the canal to a smaller one.

Turnouts can be concrete structures, or pipe structures.

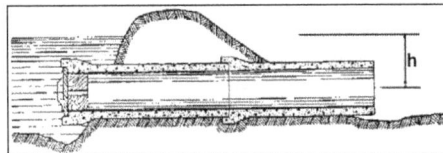

A pipe turnout.

c. Checks

To divert water from the field ditch to the field, it is often necessary to raise the water level in the ditch. Checks are structures placed across the ditch to block it temporarily and to raise the up-stream water level. Checks can be permanent structures or portable.

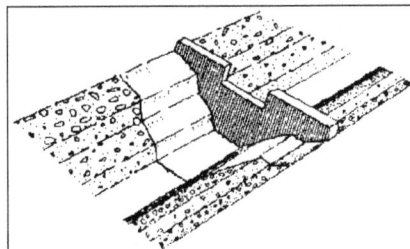

A permanent concrete check.

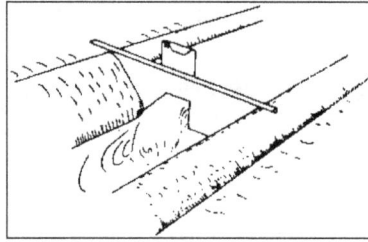
A portable metal check.

Crossing Structures

It is often necessary to carry irrigation water across roads, hillsides and natural depressions. Crossing structures, such as flumes, culverts and inverted siphons, are then required.

a. Flumes

Flumes are used to carry irrigation water across gullies, ravines or other natural depressions. They are open canals made of wood (bamboo), metal or concrete which often need to be supported by pillars.

b. Culverts

Culverts are used to carry the water across roads. The structure consists of masonry or concrete headwalls at the inlet and outlet connected by a buried pipeline.

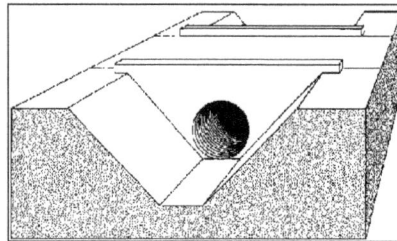
A culvert.

c. Inverted siphons

When water has to be carried across a road which is at the same level as or below the canal bottom, an inverted siphon is used instead of a culvert. The structure consists of an inlet and outlet connected by a pipeline. Inverted siphons are also used to carry water across wide depressions.

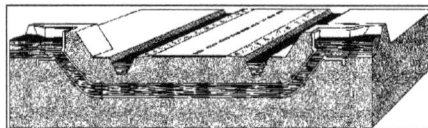
An inverted siphon.

Water Measurement Structures

The principal objective of measuring irrigation water is to permit efficient distribution and application. By measuring the flow of water, a farmer knows how much water is applied during each irrigation.

In irrigation schemes where water costs are charged to the farmer, water measurement provides a basis for estimating water charges.

The most commonly used water measuring structures are weirs and flumes. In these structures, the water depth is read on a scale which is part of the structure. Using this reading, the flow-rate is then computed from standard formulas or obtained from standard tables prepared specially for the structure.

a. Weirs

In its simplest form, a weir consists of a wall of timber, metal or concrete with an opening with fixed dimensions cut in its edge. The opening, called a notch, may be rectangular, trapezoidal or triangular.

Some examples of weirs:

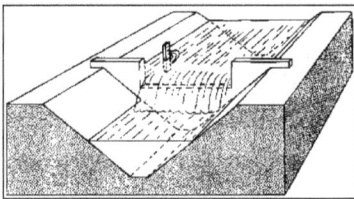

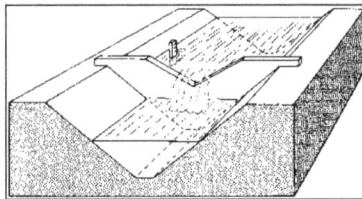

A Rectangular Weir.   A Triangular Weir.   A Trapezoidal Weir.

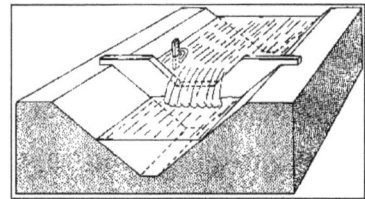

b. Parshall flumes

The Parshall flume consists of a metal or concrete channel structure with three main sections: (1) a converging section at the upstream end, leading to (2) a constricted or throat section and (3) a diverging section at the downstream end.

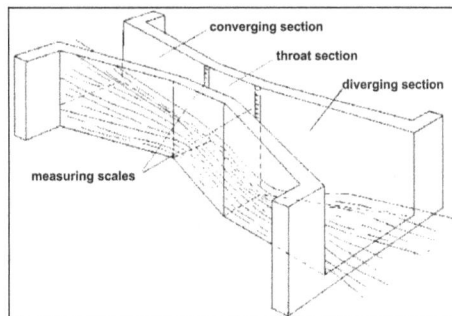

A Parshall flume.

Depending on the flow condition (free flow or submerged flow), the water depth readings are taken on one scale only (the upstream one) or on both scales simultaneously.

c. Cut-throat flume

The cut-throat flume is similar to the Parshall flume, but has no throat section, only converging and diverging sections. Unlike the Parshall flume, the cut-throat flume has a flat bottom. Because it is easier to construct and install, the cut-throat flume is often preferred to the Parshall flume.

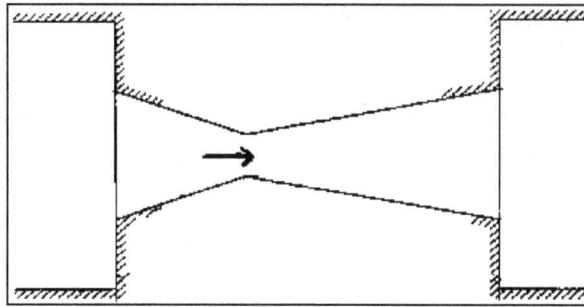
A cut-throat flume.

Drainage System

A drainage system is necessary to remove excess water from the irrigated land. This excess water may be e.g. waste water from irrigation or surface runoff from rainfall. It may also include leakage or seepage water from the distribution system.

Excess surface water is removed through shallow open drains. Excess groundwater is removed through deep open drains or underground pipes.

Factors Affecting any Selected Method of Irrigation

The following factors affect any selected method of irrigation:

Topography

Topography is the slope of the ground and how much uneven or leveled it is. The irrigation method is selected accordingly.

If the slope is from 0.4 to 8 per cent, corrugation method of irrigation is suitable. If the slope is more than the above limit, then sprinkler method has been considered more suitable for soils which are shallow and permeability is fast.

Similarly, leveled surface flow method, check basin method, furrow irrigation method etc., are adopted. Apart from level differences, the structure and composition of soil also affects the selection of irrigation method. Deep soils have medium and fast permeability for which corrugated method of irrigation is suitable. Rate of seepage of water in the soil depends on its composition.

Climate

Ground level irrigation methods are directly affected by climate. The sprinkler system of irrigation is most suitable for such climate. Dryness, humidity and speed of air affect the methods of irrigation.

Means of Irrigation

The sources of irrigation and the chemical composition of the water also affect the irrigation method. At the time of irrigation by tube wells, the sprinkler method or drip irrigation method is suitable, but in canal irrigation, flood method is selected.

Crops

The irrigation method is selected according to type of crops and pattern of its sowing. Different crops require different quantities of water. Growth of plants and their height are also affected. In taller plants, check basin irrigation method is more suitable as compared to sprinkler method.

Texture of the soils crusting, cracking and infiltration characteristics of the surface soil; nature and depth of relatively impermeable layer(s) in the sub-soil, if any; the water storage capacity of the potential root zone; the magnitude and nature of land slope; the length and size of the field; surface drainage; the nature and extent of salts in the surface soil and the sub-soil, also decide the method of irrigation.

Water Conservation

Water conservation is the demand of the day when the whole world is facing water crisis. Hence, such a method of irrigation should be adopted which uses minimum water but provides maximum humidity to the plants. Sprinkler method and drip irrigation method are the best from the view point of water conservation.

Economic Factors and Labour

While selecting any method of irrigation, the economic condition must be kept in mind as many irrigation methods require heavy initial investment but lesser investment later on. Surface irrigation methods require continuous investment. Along with investment, availability of labour is also an essential factor. As compared to sprinkler method, check basin method of irrigation requires more labour.

Outlook, management skills and financial resources of the farmer; nature of farm machinery used; availability and cost of labour; maintenance facilities and costs of irrigation equipment; availability of power supply should also be considered.

As far as possible, an irrigation method should not only provide a high level of water application efficiency but also ensure its economic viability, sustained soil productivity and wide adaptability to the prevalent features of the farm.

Chapter 2

Fundamental Aspects of Irrigation Engineering

The branch of engineering which focuses on the analysis and design of irrigation systems such as barrages, canals, drains and dams is known as irrigation engineering. Some of the important aspects of this field are planning irrigation projects, assessing charges for irrigation water and selecting irrigation methods. The chapter closely examines these key aspects of irrigation engineering to provide an extensive understanding of the subject.

Irrigation Engineering is important since it helps determine future Irrigation expectations. Irrigation has been a central feature of agriculture for over 5000 years, and was the basis of the economy and society of numerous societies, ranging from Asia to Arizona. Irragation can be termed as the artificial process of applying water to the soil to help in growing agricultural crops or maintaining the landscapes when there is shortage of natural water by rain. Additionally, irrigation also has a few other uses in crop production, which include protecting plants against frost, suppressing weed growth in grain fields and preventing soil consolidation. In contrast, agriculture that relies only on direct rainfall is referred to as rain-fed or dryland farming. Irrigation systems are also used for dust suppression, disposal of sewage, and in mining. Irrigation is often studied together with drainage, which is the natural or artificial removal of surface and sub-surface water from a given area. water is required for agriculture. sometimes this water requirement is fulfilled by rain, but there are some dry areas where irrigation is the only process by which water is supplied to crops.

Irrigation engineering deals with the analysis and design of irrigation systems which include dams, weir, barrage, canals, drains and other supporting systems etc. Good knowledge of hydraulics or fluid mechanics is very much required for design of irrigation system.

Planning of an Irrigation Project

An irrigation project is an agricultural establishment which can supply controlled amounts of water to lands for growing crops. Irrigation projects mainly consist of hydraulic structures which collect (from a source), convey and deliver (to farm fields) water for irrigation. A small irrigation project may consist of a small diversion weir (or a pumping plant) with small channels and some minor control structures.

A large irrigation project includes a large storage reservoir created by a huge dam (or a long weir or barrage), hundreds of kilometres of canals, branches and distributaries, control structures and other works. Large irrigation projects are generally built as multipurpose projects which also serve to generate hydropower, control floods, and meet water supply and other demands.

Assuming other factors (such as enlightened farmer, availability of good infrastructure for supplying input material and marketing the agricultural production) reasonably favourable, the following are the essential conditions for the success of any irrigation project:

- Suitability of land (with respect to its soil, topography and drainage features) for the purpose of agriculture.

- Favourable climatic conditions for proper growth and yield of the crops.

- Adequate and economic supply of suitable quality of water.

- Good site condition for the construction and operation of engineering works.

A small irrigation project can be developed in a relatively short time. For example, a farmer can develop his own tube well irrigation system by securing bank loan and, soon after, getting the engineering works constructed. However, development of a large irrigation project is more complicated and time-consuming due to the associated organisational, financial, legal, administrative, environmental and engineering problems.

The main stages of a large irrigation project are:

- The promotional stage,

- The planning stage,

- The construction stage, and

- The settlement stage.

The planning stage itself consists of three sub-stages:

- Preliminary planning, including feasibility studies,

- Detailed planning of water and land use, and

- The design of irrigation structures and canals.

The feasibility of an irrigation project is decided on the basis of preliminary estimates of:

- Area of land suitable for irrigation,

- Water requirements,

- Available water supplies,

- Productivity of irrigated land, and

- Required engineering works.

An irrigation project is considered feasible if the total estimated benefits of the project exceed its total estimated cost. Adequate planning of all aspects (organisational, technical, agricultural, legal, environmental and financial) is always essential for a feasible irrigation project.

The process of planning of an irrigation project is divided into the following two stages:

- Preliminary planning, and

- Detailed planning.

Preliminary plans, usually based on available information, are generally approximate, but set the course for detailed planning. Based on preliminary planning, the accurate measurements are taken and, thus, more accurate detailed plans are prepared. The detailed plans may, however, have to be altered at different stages of the project.

The following are the main factors which must be determined accurately during the planning stage of an irrigation project:

- Type of project and general plan of irrigation works.
- Location, extent and type of irrigable lands.
- Irrigation requirements for profitable crop production.
- Available water supplies for the project.
- Culturable areas which can be economically supplied with water.
- Types and locations of necessary engineering works.
- Needs for immediate and future drainage.
- Feasibility of hydroelectric power development.
- Cost of storage, irrigation, power and drainage features.
- Evaluation of probable power, income and indirect benefits.
- Method of financing the project construction.
- Desirable type of construction and development.
- Probable annual cost of water to the farmers.
- Cost of land preparations and farm distribution systems.
- Feasible crops, costs of crop production and probable crop returns.

The preliminary planning of an irrigation project consists of collecting and analysing all available data for the purpose, securing additional data needed by limited field surveys and determining the feasibility of the proposed development by making preliminary study of major features in sufficient detail.

For detailed planning, accurate data on all aspects of the proposed irrigation project are needed to prepare plans and designs of various components, and also to determine their most suitable site locations. There can be different feasible plans and designs possible for a particular project. Merits and demerits of all such possible alternatives must be looked into before arriving at a final plan for the project.

Financial Analysis of Irrigation Project

Analyzing the financial benefits of an irrigation project involves looking at the two levels: - i) farmer level and the ii) scheme level. At farmer level, we look at production levels, labour requirements and net income 'with' and 'without' the project. At scheme level, we look at costs incurred in constructing, operating and managing the whole scheme.

Financial Analysis of an irrigation project consists of the following:-

Farm Income Analysis

During project analysis, the underlying assumption made is that for farming community or for a farm, the objective will be maximization of income that the families will earn as a result of the participation in the project. The resources used are land, water, electricity and labour. The tools to evaluate these resources are: -

a. Cropping Patterns: When an irrigation project is introduced, the area of irrigation comes from the participating farmers' landholdings being used for rainfed cultivation.

    If the farmers become full time irrigators, this will mean that income from cultivation is lost and income from irrigation is gained. In order to assess the impact of this, cropping patterns are analyzed and suitable decisions are taken.

b. Labour Requirements: Labour requirements are calculated on crop by crop basis and added to estimate the total requirements in any given situation. Where an exhaustive survey on the labour schemes has been carried out, this provides the data associated with various operations in the proposed scheme. When calculating the requirements for each crop, not only the total requirements but also the distribution over the cropping period will have to be established so that labour requirements in the peak periods can be determined.

c. Crop Budgets/Gross Margin Analysis: Crop Budgets contain the evaluation of gross margins per hectare for the different crops. Gross margin is the income generated from a production activity and is equal to the difference between the total gross income and the total variable costs.

Scheme Investment Analysis

The scheme investment analysis looks at the scheme income based on the gross margins, investment costs and the operations and maintenance costs. The analysis seeks to judge the likely incremental benefits project participants and the incentive for farmers to participate in the project, thus looking at the attractiveness of the project to the indulging farmers.

Scheme investment analysis depends on the following:-

a. Investment: Investment refers to the amount put in a project irrespective of its type. Investment can also be termed as initial costs incurred to kick start a project.

b. Land: Land for any irrigation project must appear in the investment analysis of the project. Similarly, rent should also appear as a cost in the investment analysis if the land has been rented.

c. Operating Costs: The operating expenditure is calculated for the costs of equipment utilized in making the investment work functional and would include;

    •   Replacement Costs: These are the costs incurred to replace specific items.

- Energy Costs: This depends on the elevation of the water source relative to the elevation of the scheme, which determines whether water should be pumped in order to reach the scheme and the irrigation system used.

- Repair and Maintenance Costs: These costs are usually assumed to depend on the cost of the equipment utilized. Thus a percentage of the cost of the equipment (generally between 1.5-5%) is taken as the repair and maintenance costs per year.

- Water Charges: These are the charges payable to whoever supplies the water, for example the national water authority. Where water is purchased, the water charges should be indicated as a cost.

   d. Other Costs: Following come under the other costs;

- Sunk Cost: A cost incurred in the past projects that cannot be recovered again.

- Residual Value: This is the value of the asset remaining unused at the end of a project. The asset can be termed as an residual asset.

Setting up the Investment Budget

Having assessed the costs and benefits, the budget of the irrigation project is set up.

Project Period

This is defined as the time duration for which the project will the carried out. If the project centers on only one major asset, say the irrigation system, then the usual project period is said to be 20 years. For external funding, the projects get wrapped up by 5-7 years.

Time Value of Money

When costs and benefits are spread over time, then the future income has to be reduced to its present worth. This is based on the principle that a dollar buys more today than it will buy tomorrow.

Determining the discount rate: The factor used to reduce projected future income or to accumulate loans now, is the discount rate or interest. There are two main explanations for interest namely, time preference and opportunity cost of capital.

Measuring the Project Worthiness

The viability or worthiness of the project that takes the timing of costs and benefits into account can be measured using the following indicators;

a. Net Present Value (NPV): In financial terms, the net present value (NPV) or net present worth of a time series of cash flows, both incoming and outgoing, is defined as the sum of the present values (PVs) of the individual cash flows of the same entity. In the case when all future cash flows are incoming (such as coupons and principal of a bond) and the only outflow of cash is the purchase price, the NPV is simply the PV of future cash flows minus the purchase price (which is its own PV). NPV is a central tool in discounted cash flow (DCF) analysis and is a standard method for using the time value of money to appraise long-term projects. Used for capital budgeting and widely used throughout economics, finance, and accounting, it measures the excess or shortfall of cash flows, in present value terms, above the cost of funds.

Formula for NPV is

$$\frac{R_t}{(1+i)^t}$$

where,

>   t – time of the cash flow,

>   i – discount rate (the rate of return that could be earned on an investment in the financial markets with similar risk.),

> R_t – the net cash flow i.e. cash inflow – cash outflow, at time t. For educational purposes,

> R_0 - is commonly placed to the left of the sum to emphasize its role as (minus) the investment.

The result of this formula is multiplied with the Annual Net cash in-flows and reduced by Initial Cash outlay the present value but in cases where the cash flows are not equal in amount, then the previous formula will be used to determine the present value of each cash flow separately. Any cash flow within 12 months will not be discounted for NPV purpose, nevertheless the usual initial investments during the first year R0 are summed up a negative cash flow.

b. Benefit/Cost Ratio: A benefit-cost ratio (BCR) is an indicator, used in the formal discipline of cost-benefit analysis, that attempts to summarize the overall value for money of a project or proposal. A BCR is the ratio of the benefits of a project or proposal, expressed in monetary terms, relative to its costs, also expressed in monetary terms. All benefits and costs should be expressed in discounted present values. Benefit cost ratio (BCR) takes into account the amount of monetary gain realized by performing a project versus the amount it costs to execute the project. The higher the BCR the better the investment. General rule of thumb is that if the benefit is higher than the cost the project is a good investment.

c. Internal Rate of Return: The internal rate of return on an investment or project is the "annualized effective compounded return rate" or "rate of return" that makes the net present value (NPV as NET*1/(1+IRR)^year) of all cash flows (both positive and negative) from a particular investment equal to zero. It can also be defined as the discount rate at which the present value of all future cash flow is equal to the initial investment or in other words the rate at which an investment breaks even.

In more specific terms, the IRR of an investment is the discount rate at which the net present value of costs (negative cash flows) of the investment equals the net present value of the benefits (positive cash flows) of the investment.

IRR calculations are commonly used to evaluate the desirability of investments or projects. The higher a project's IRR, the more desirable it is to undertake the project. Assuming all projects require the same amount of up-front investment, the project with the highest IRR would be considered the best and undertaken first.

A firm (or individual) should, in theory, undertake all projects or investments available with IRRs that exceed the cost of capital. Investment may be limited by availability of funds to the firm and/or by the firm's capacity or ability to manage numerous projects.

Given a collection of pairs (time, cash flow) involved in a project, the internal rate of return follows from the net present value as a function of the rate of return. A rate of return for which this function is zero is an internal rate of return.

Given the (period, cash flow) pairs (η, C_n) where is a positive integer, the total number of periods, and the net present value , the internal rate of return is given by in:

$$\text{NPV} = \sum_{n=0}^{N} \frac{C_n}{(1+r)^n} = 0$$

The period is usually given in years, but the calculation may be made simpler if is calculated using the period in which the majority of the problem is defined (e.g., using months if most of the cash flows occur at monthly intervals) and converted to a yearly period thereafter.

Any fixed time can be used in place of the present (e.g., the end of one interval of an annuity); the value obtained is zero if and only if the NPV is zero.

In the case that the cash flows are random variables, such as in the case of a life annuity, the expected values are put into the above formula.

Often, the value of cannot be found analytically. In this case, numerical methods or graphical methods must be used.

Irrigation Project Costs

Irrigation project costs include all the expenditure made to establish, maintain and operate a project. Costs are estimated on an annual basis. The annual cost of a project includes both fixed and variable costs.

Fixed Costs

Fixed costs, also referred as investment or initial costs, include the following, as applicable:

   (a) Costs of obtaining water right and permits.

   (b) Planning and design costs.

   (c) Land purchase and rehabilitation of the population in the areas affected by the water resources project.

   (d) Cost of storage reservoirs, head regulator and canal water distribution system, including associated structures and controls.

(e) Command area development surveying, land development operations and on-farm water conveyance and control.

(f) Drainage system main drains, link drains, and no-farm drainage system surface and sub-surface.

(g) Cost of wells, pumps, electric motors/engines and pumping plant accessories, and their installation.

(h) Pump house

(i) Electric Power connection, metering and recording equipment.

(j) Automation equipment/remote control, if used.

(k) Inspection and approach roads.

(l) Equipment for water application, sprinkler/drip irrigation equipment, if used.

The above costs are incurred at the initial stages of the project, while others are paid annually. Annual fixed costs include the interest on the total investment on the project.

Variable Costs

Variable costs are recurring in nature and computed on an annual basis. They are operation and maintenance costs as well as levies and charges on insurance and miscellaneous operating costs of recurring nature. The variable annual costs may be enumerated as follows:

(a) Maintenance of structures and water distribution network.

(b) Costs of fuel, namely, diesel or other fuels and electricity.

(c) Lubricants, minor repairs, and painting.

(d) Layout of field for surface irrigation renewal of borders, ridges and field channels.

(e) Maintenance of drainage system desilting, weed control etc.

(f) Operating manpower cost (Manpower costs include salaries, social benefits, housing, insurance, medical treatment, transportation and similar items). A simple procedure, commonly used for preliminary cost estimates is to calculate the interest on the average value of the installation at the prevailing interest rate:

$$\text{Annual interest cost} = \frac{(\text{Value of installation - salvage value}) \times \text{Interest rate}}{2}$$

Depreciation

It is a provision of funds over the life time of the project for its replacement. Depreciation is excluded from the economic appraisal of a project as it is only an accounting concept. Depreciation is the anticipated reduction in the value of an asset due to physical use of the equipment/

structure or obsolescence. In the conventional analysis, the annual depreciation is computed as follows:

$$\text{Annual depreciation} = \frac{\text{Original cost - salvage value}}{\text{Useful life in years}}$$

Depreciation refers to the process of allocating a portion of the original cost of a fixed asset to each accounting period so that the value is gradually written off during the course of the estimated useful life of the asset. Allowance may be made for the asset's estimated resale value, if any, at the end of the useful life of the enterprise.

In discounted cash flow analysis, depreciation is not treated as a cost. The cost of an asset is shown in the year it is incurred and the benefits are shown in the year they are obtained. Since this is done over the entire life of the project, it is not necessary to show the value of the asset apportioned in any given year as depreciation. That would amount to double counting.

Service Period of Wells and Pumps Projects

In the case of ground water and lift irrigation projects, when the expected service period of wells and pumps is specified in hours of operation, same in years is calculated by dividing the total hours of operation by the average annual hours of operation by the average annual hours of operation.

Variable Costs of Irrigation Projects

If the electrical connection charges are paid in lump sum, the annualized cost may be estimated, assuming an expected service life of about 25 years. If the charges are to be paid annually, the same are to be added to the operation and maintenance cost, to arrive at the annual variable cost.

Variable costs of surface irrigation projects include the costs on regulation of the conveyance system as well as maintenance and repairs. In case of wells and pumps, they include the cost of power/fuel (electricity/diesel), cost of lubricants, labour charges for operating the pumping units and the expenditure on repairs and maintenance of the equipment and accessories.

The cost of power is often the most important component of variable costs in the case of pumping systems. The usual practice is to operation from the known discharge rate of the pumping plant, total operating head and its overall efficiency. The requirement of power is expressed in kilowatt-hour per hour for electricity and liters of diesel per hour of operation of engines.

The energy consumption of an electric motor is computed as follows:

$$\text{Energy Consumption} = \frac{\text{Brake horse Power}}{\text{Motor efficiency}} \times 0.746$$

Efficiencies of electric motor may be obtained from the performance data supplied by the manufacturers. Motor efficiencies usually vary from 75 to 90 per cent.

The demand of electrical power for hourly operation is multiplied by the annual hours of operation to arrive at the total annual energy consumption. The annual power cost is determined by multiplying the annual energy demand by the prevailing cost per unit of electrical energy.

In case of engine, the cost of fuel is computed as follows:

Fuel cost = BHP× Specific fuel consumption ×Cost of fuel per liter

A realistic estimate of the rate of fuel consumption for a given engine can be made if the manufacturer's fuel consumption curve for the engine is available. Fuel consumption of diesel commonly used in irrigation pumping vary from 0.2 L to 0.29 L per bhp hour. An average value of 0.23 L/bhp-h can be assumed in the absence of better data.

The consumption of lubrication oil is usually assumed to be 4.5 L per 1000 bhp-h. Many manufacturers provide values for the consumption of lubricants of their products. From the cost of lubricants per hour of operation, the annual cost of lubricants is computed.

Types of Canal Construction Machineries

Construction of canals requires different types of machineries and equipments, the selection of which depends on many factors. These types of canal machines and their selection is discussed.

Canals are substantially significant infrastructures that considerably affect the economy of the county or region in many ways. Therefore, it is necessary to construct such sizable project in the best possible way with lowest budget.

Canal Construction Equipment.

Canal Construction Machines.

So, the selection of construction machineries directly influences the cost of the project because not only does it decrease required man power but also improve the quality of the project and decline the construction period.

In order to specify the best and most efficient machineries, engineers should be familiar with the machine capacity, their suitability to his/her project, and other machinery aspects.

Factors Considered for Selection of Canal Construction Machineries

Different types of machines are required for different canal construction. Following factors are considered for their selection:

- Top width of the canal.
- Bottom width of the canal.
- Percentage of slope on the walls.
- Whether the slope is constant throughout the structure.
- Number of variations in the profiles.
- The existence of a haul road.
- Expected concrete supply to the paver.

Types of Canal Construction Machineries

- Canal excavators
- Canal Liners
- Loaders
- Mobile cranes
- Towable backhoe
- Canal finisher
- Crusher
- Dozer
- Dumpers
- Canal paver
- Concrete mixer.

Canal Excavators

There are different types of excavation machines for digging canals such as bucket wheel excavator and hinge bucket chain trimmer. The former can be used to excavate canals with up to 9m top breadth. The latter can dig canals with up to 25m top widths.

In addition to full section canals, the hinge bucker trimmer machine can be used for half section canal, extended half section canal and separated by berms.

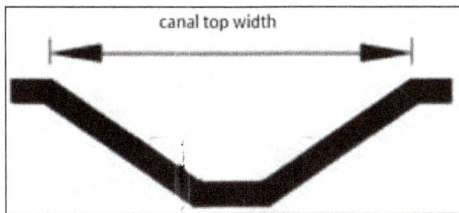

Top Width of Canal.

Bucket Wheel Excavator.

Bucket Chain Excavator.

Half Canal Cross Section Trimmer Machine.

Canal Liners

Lining of canals is crucially significant because they eliminate seepage in the canal and consequently the loss of water would be avoided. Various machine liners with various capacities and sizes are available for canal lining such as towed boat liner, telescopic liner and hinge concrete liner.

Towed boat liner can be used for canals with 6m top width and capable of lining up to 60m^2 per hour. Telescopic liner is capable of handling canals with 9m top width and the lining rate is 180m^2 per hour. As far as hinge concrete liner is concerned, it used for canals with top breadth of up to 25m. Similar to hinge bucket chain trimmer, hinge concrete liner may be used for full section canals, half section canals, extended half section canals and separated by berms. Figure shows hinge concrete liner machine.

Towed Boat Liner.

Telescopic Liner.

Hinge Concrete Liner.

Loaders

Loaders are used to clean construction site from debris, mud and excavated materials and upload these materials into to other vehicles to transport such materials away from construction site. It also may be used to transport dry construction materials in project site.

There are different types of loader such as wheel loader, backhoe loader, skip loader and telescopic loaders. Figure show different loader machine respectively.

Wheel Loader.

Backhoe Loader.

Skip Loader.

Telescopic Loader.

Mobile Cranes

Cranes are machineries similar to towers which consist of base, jib, motor and mast. Mobile cranes are commonly mounted on vehicles and used to lift or down materials.

Mobile Cranes.

Towable Backhoe

It is small size machinery which can be used in the construction of canals for many purposes such as clearing construction site and transporting light weight construction materials to the project site.

Towable backhoe has a bucket in front and toe at the back and both can be replaced.

Towable Backhoe.

Canal Finisher

Finishing canals is very important because it would avoid water logging. Finishing is conducted through concrete lining on both sides of the canal.

Vehicles that carry concrete can feed canal finisher from one side only and a conveyor which spans the canal, used to transport the concrete to the other side of the canal. There are transverse and longitudinal joints created by attachments on the canal finisher. The joints will be finished by worker on the work bridge finish joint.

Full Canal Cross Section Finisher.

Half Cross-Section Canal Finisher.

Crusher

It is used to crush rocks and stone to smaller sizes and there are various types of crusher which can be used in the canal construction. For example, impact crusher, roller crusher, jaw crusher, cone crusher and pressure crusher.

Impact Crusher.

Dozers

Dozers are complicated to drive that has continuous treads and a board hydraulic blade in front. The types of the blade used controls the function of the dozer. It is applied for many purposes for example earth removing, cleaning and grading lands.

Bulldozer.

Dumpers

Dumpers are small size machines which uses diesel to generate power. There are two main types of dumpers categorized based on the machine size such as dumper trucks and dumper crawler.

The main functions of dumper in canal construction are clearing construction site and transport construction materials to the project site.

Dumper Trucks.

Crawler Dumper.

Canal Paver Machine

Paving canal is an alternative to lining canal, so either lining or paving is conducted for the canal. Paving machine used to pave the canal and it covers both sides of the canal.

The width of finishing is quite large, so the paving operation can be carried out in a higher rate. The machine can be configured to pave canals with different slope breadth.

Canal Paver Machine Followed by Stop Water Insertion Machine.

Concrete Mixer

It is well known machinery that used to mix concrete components to achieve uniform mixture. There are mobile concrete mixer and fixed concrete mixture.

Mobile Concrete Mixer.

Stationary Concrete Mixer.

Assessment of Charges for Irrigation Water

Assessment of irrigation water charges can be done in one of the following ways:

- Assessment on area basis.

- Volumetric assessment.

- Assessment based on outlet capacity.

- Permanent assessment.

- Consolidated assessment.

In the area basis method of assessment, water charges are fixed per unit area of land irrigated for each of the crops grown. The rates of water charges depend on the cash value of crop, water requirement of crop and the time of water demand with respect to the available supplies in the source.

Since the water charges are not related to the actual quantity of water used, the farmers (particularly those whose holdings are in the head reaches of the canal) tend to over irrigate their land. This results in uneconomical use of available irrigation water beside depriving the cultivators in the tail reaches of the canal of their due share of irrigation water.

However, this method of assessment, being simple and convenient, is generally used for almost all irrigation projects. In the volumetric assessment, the charges are in proportion to the actual amount of water received by the cultivator. This method, therefore, requires installation of water meters at all the outlets of the irrigation system.

Alternatively, modular outlets may be provided to supply a specified discharge of water. This method results in economical use of irrigation water and is, therefore, an ideal method of assessment. But it has several drawbacks.

This method requires installation and maintenance of suitable devices for water measurement. These devices require adequate head at the outlet. Further, there is a possibility of water theft by cutting of banks or siphoning over the bank through a flexible hose pipe. Also, the distribution of charges among the farmers, whose holdings are served by a common outlet, may be difficult. Because of these drawbacks.

Assessment of canal water charges based on outlet capacity is a simple method and is workable if the outlets are rigid or semi-modular and the channel may run within outlet's designed range of parameters. In some regions, artificial irrigation, though not essential, has been provided to meet the water demand only in drought years.

Every farmer of such a region has to pay a fixed amount. The farmers have to pay these charges even for the years in which they do not take any water. A farmer has also to pay tax on the land owned by him. In the consolidated assessment method, both the land revenue and the water charges are combined, and the cultivators are accordingly charged.

Irrigation Efficiency

The term irrigation efficiency expresses the performance of a complete irrigation system or components of the system. Irrigation efficiency is defined as the ratio between the amount of water used to meet the consumptive use requirement of crop plus that necessary to maintain a favourable salt balance in the crop root zone to the total volume of water diverted, stored or pumped for irrigation. Thus, water applied by the irrigation system and not being made available to be taken up by plant roots is wasted and reduces irrigation efficiency. Figure shows components of water loss from source to point of application. In addition, losses can also occur during storage in case of pond, tank, or reservoirs. The major causes for reduced irrigation efficiency include storage losses, conveyance losses and field application losses. Overall irrigation efficiency of major irrigation projects ranges between 35-40%. This is one of the reasons for increasing gap between irrigation potential created (102.77 M ha till end of 10th plan 2007) and utilized (87.23 M ha). This gap of about 16%, is same as the irrigation potential created between 1951 and 1970. At the end of eighth plan, Planning commission estimated that with a 10% increase in the present level of water use efficiency in irrigation systems, an additional 14 Mha area can be brought under irrigation from the existing irrigation capacities. In order to meet the growing demands of water for food, environment, urban and industry, it is necessary to improve irrigation efficiency at all levels.

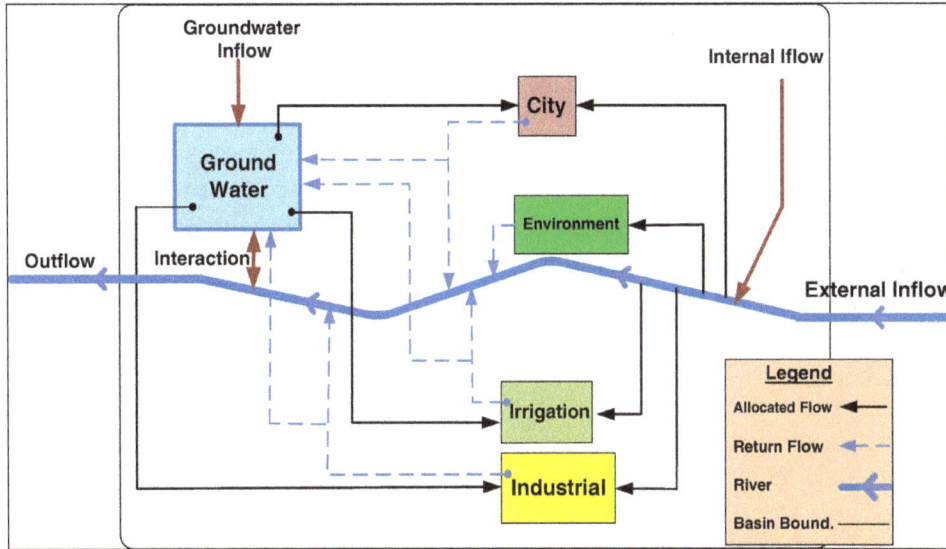

Definition of Various Efficiencies

Reservoir Storage Efficiency

It is the efficiency with which water is stored in the reservoir. It is expressed as follows,

$$E_r = 100 \times \left(1 - \frac{V_e + V_s}{V_i} \right) = 100 \left(\frac{V_o + \Delta S}{V_i} \right)$$

Where,

V_e = evaporation volume from the reservoir

V_s = seepage volume from the reservoir

V_t = inflow to the reservoir

V_o = volume of out flow from the reservoir

ΔS = change in reservoir storage.

Water Conveyance Efficiency

The conveyance efficiency is used to measure the efficiency of water conveyance systems associated with the canal network, water courses and field channels. It isdefined as the ratio between the water that reaches a farm or field and that diverted from the irrigation water source. Mathematically it is represented as follows:

$$E_c = 100(V_f / V_d)$$

Where,

E_c = the conveyance efficiency (%),

V_f = the volume of water that reaches the farm or field (m³),

V_d = the volume of water diverted (m³) from the source.

E_c also applies to segments of canals or pipelines, where the water losses include canal seepage or leaks in pipelines. The global E_c can be computed as the product of the individual component efficiencies, E_{ci}, where i representthe segment number. Typically, conveyance losses are much lower for closed conduits or pipelines compared with unlined or lined canals. Even theconveyance efficiency of lined canals may decline over time due to material deterioration or poor maintenance.

Application Efficiency

Application efficiency relates to the actual storage of water in the root zone to meet the crop water needs in relation to the water applied to the field. It might be defined for individual irrigation or parts of irrigations or irrigation sets.Application efficiency includes any application losses to evaporation or seepage from surface water channels or furrows, any leaks from sprinkler or drip pipelines, percolation beneath the root zone, drift from sprinklers, evaporation of droplets in the air, or runoff from the field. In case of surface irrigation evaporation losses are generally small but runoff and deep percolation are substantial. However, air losses (droplet evaporation and drift) can be very large if the sprinkler design or excessive pressure produces a high percentage of very fine droplets. Application efficiency is defined as:

$$E_a = 100(V_s / V_f)$$

Where,

E_a = the application efficiency (%),

V_s = the volume of water stored in root zone (m³),

V_f = the water delivered to the field or farm (m³).

Table: Typical values of application efficiency for different irrigation systems.

System Type	Application Efficiency Range (%)
Surface Irrigation	
Basin	60-95
Border	60-90
Furrow	50-90
Surge	60-90

Sprinkler Irrigation	
Handmove	65-80
Traveling Gun	60-70
Center Pivot & Linear	70-95
Solid Set	70-85
Microirrigation	
Point source emitters	75-95
Line source emitter	70-95

Storage Efficiency

The water storage efficiency evaluates the storage of water in the root zone after the irrigation in relation to the need of water prior to irrigation:

$$E_s = 100 \, (V_s / V_{rz})$$

Where,

E_s = the storage efficiency (%)

V_{rz} = the root zone storage capacity (m³).

The root zone depth and the water-holding capacity of the root zone determine V_{rz}. The storage efficiency has little utility for sprinkler ormicro irrigation because these irrigation methods seldom completely refill the root zone.

Water Distribution Efficiency

It is the ratio between the mean of numerical deviations from the average depth of water stored-during irrigation (Y) and the average depth stored during irrigation (d). It is mathematically expressed as:

$$E_d = \left(1 - \frac{\bar{y}}{\bar{d}}\right) \times 100$$

Where,

Y = Average numerical deviation in depth of water stored from average depthstored during irrigation.

d = Average depth of water stored during irrigation.

It is a measure of water distribution within the field. A low distribution efficiencymeans non-uniformity in the distribution of irrigation water. This may be due to uneven landlevelling. There may beexisting low patcheswhere water will penetrate more and high patches where water cannot reach. This leaves some spots unirrigated unless excess irrigation water is applied. Excess waterapplication lowers irrigation efficiency.It may be noted that water distribution efficiency is identical to Christiansen's Uniformity Coefficient.

Water use Efficiency

The term water use efficiency denotes the production of crops per unit water applied. It is expressed as the weight of crop produce per unit depth of water over a unit area. i.e., kg/cm/ha.

Crop Water use Efficiency

It is the ratio of crop yield per amount of water depleted by the crop in the process of evapotranspiration (ET).

Crop water use efficiency = Y/ET

Field Water use Efficiency

It is the ratio of crop yield (Y) to the total amount of water used in the field (WR).

Field water use efficiency = Y/WR

Irrigation Uniformity

Uniformity is a measure to describe evenness ofwater application over the length of the field.It is a statistical measure of the distribution of the applied water, which is affected by various factors suchas the method of irrigation, topography, infiltrationcharacteristics, and hydraulic characteristics(pressure, flow rate, etc.) of the irrigation system.

It is generally expressed using Christiansen'scoefficient of uniformity (CU), distribution uniformity(DU) and emission uniformity (EU) for drip irrigationsystems.Irrigation application distributions are usually based on depths of water (volume per unit area); however, for micro irrigation systems they are usually based on emitter flow volumes because the entire land area is not typically wetted.

Christiansen's Uniformity Coefficient

Christiansenproposed acoefficient intended mainly for sprinkler system based on the catch volumes given as:

$$CU = \left[\frac{1 - \left(\sum |X - x| \right)}{\sum X} \right]$$

Where, CU is the Christiansen's uniformity coefficient in percent, X is the depth (or volume) of water in each of the equally spaced catch containers in mm or ml, and x is the mean depth (volume) of the catch (mm or ml).

Low-Quarter Distribution Uniformity

It is defined as the ratio of the average infiltration in the lower quarter to the average infiltration over the entire field:

$$DU = 100 \left(\frac{V_p}{V_t} \right)$$

Where,

DU is the distribution uniformity (%) for the lower quarter of the field, V_p is the mean application volume (m³)or depth in the lower quarter, and V_f is the mean application volume (m³)or depth for the whole field.

Deep Percolation Ratio and Tail Water Ratio

The Deep Percolation Ratio (DPR) and Tail Water Ratio (TWR) were developed to take into account the losses occurring via deep percolation and runoff in surface irrigation methods.

Deep Percolation Ratio

It is the ratio of amount or depth of deep percolation to the amount or depth of applied water and is expressed as:

$$DPR = \frac{V_{Dp}}{V_f} \times 100.$$

Where,

V_{DP} = amount of water lost to deep percolation,

V_f = amount of water delivered to the field.

Tail Water Ratio

It is the ratio of amount or depth of runoff to the amount or depth of applied water and is expressed as:

$$TWR = \frac{V_{ro}}{V_f} \times 100$$

Where,

V_{ro} = amount of runoff from field.

Overall Project Efficiency

It is the ratio between the average depth of water stored in the root zone duringirrigation and water diverted from the reservoir. It is mathematically expressed as:

$$E_o = \frac{V_s}{V_d} \times 100$$

Where,

E_o = overall efficiency (%)

V_s = Water stored in the root zone (cm)

V_d = Water diverted from the reservoir (cm)

Or

$$E_o = \frac{E_a}{100} \times \frac{E_c}{100}.$$

Design, Construction and Maintenance of Irrigation Reservoirs

An irrigation reservoir can provide a farm with the irrigation water supply needed. Having a secure, dependable irrigation water supply is important to maintaining production in dry years. Combining an irrigation reservoir with a low yielding well can result in sufficient water supply. Pairing an irrigation reservoir with winter/spring pumping from a stream can result in sustainable farm water supplies while avoiding stream pumping during summer low flow periods.

Irrigation reservoirs store water pumped from an approved water source for use at a later date. A well-designed, constructed and maintained irrigation reservoir project should:

- Provide adequate water storage for the projected irrigation needs.

- Provide an efficient and cost-effective operation of the irrigation system.

- Minimize water loss from storage.

- Minimize maintenance requirements.

- Minimize both maintenance and construction costs.

- Provide a safe and secure water storage system.

Types of Irrigation Reservoirs

Soil there are three types of irrigation reservoirs.

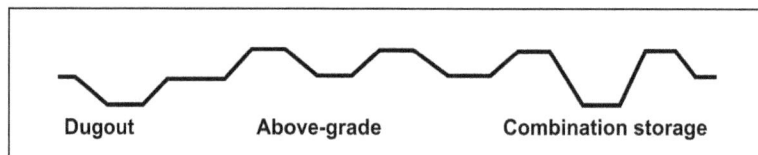

Dugout   Above-grade   Combination storage

Three types of irrigation reservoirs.

Dugout storage is below grade (normal ground level).

Large amounts of soil is excavated and disposed of during construction.

Example of a below grade reservoir or dugout.

Above grade bermed storage consists of four-sided berms constructed completely above ground to retain water. A large quantity of clay (impermeable) soil is required to build the berms and construction is similar to the construction of dams. This reservoir type is primarily used in areas with poor subsoil conditions such as bedrock.

Example of an above grade reservoir.

Combination storage is both below and above grade. The excavated soil is used to construct berms around the excavation for additional above-grade water storage. This type of construction is usually the most efficient useof the excavated material.

Example of an above and below grade reservoir or combination storage.

Storage Size

Several factors affect the volume of storage required:

- Area of the crop being irrigated.

- Type of crop being irrigated.

- The expected number and frequency of irrigations.

- How often the reservoir can be refilled.

Table: Example reservoir dimensions and water holding capacities.

Length (m)	Width (m)	Berm Height (m)	Total Depth (m)	Reservoir Footprint (m²)	Reservoir Footprint (acres)	Water Volume (m³)	Water Volume (Acre-inches)
100	100	3	8	15,376	3.8	40,122	390
100	50	3	8	9,176	2.3	13,626	133
50	50	3	8	5,476	1.4	5,130	50
50	50	1.5	5	4,225	1.0	5,324	52
50	30	1	4.5	2,604	0.6	2,356	23

Table: Example reservoir dimensions and water holding capacities showing impact of varying slope.

Length (m)	Width (m)	Berm Height (m)	Total Depth (m)	Interior Slopes	Exterior Slopes	Reservoir Footprint (m²)	Reservoir Footprint (acres)	Water Volume (m3)	Water Volume (Acre-inches)
100	100	3	8	3:1	3:1	15,376	3.8	40,122	390
100	100	3	8	6:1	4:1	16,900	4.2	20,519	200
50	50	3	8	3:1	2.5:1	5,041	1.2	5,130	50
50	50	3	8	3:1	3:1	5,476	1.4	5,130	50

Once the required storage volume is determined, calculate various reservoir layouts based on a range of water surface areas and depths. Remember, the shallower the reservoir, the larger the surface area and the more land area used for the project.

A reservoir with 3:1 inside slopes and 3:1 outside slopes; 3 m top of berm width; 10% of reservoir depth is reserved for freeboard.

Location

Locate the reservoir:

- Close to the water source that will be used to fill it, to minimize pumping costs.

- In a central location to the fields being irrigated, to minimize piping and pumping costs.

- Away from homes, buildings and public roads so that public and local infrastructure is protected from damage in the case of a major failure.

When selecting a site for the reservoir, consider the following:

- If feasible, ensure the availability of electrical power.

- Ensure proximity to access roads.

- Identify and avoid any areas with utilities, communication lines, pipelines, etc.

- Try to select a site with low-productivity land.

- Do not interfere with existing crop and field management practices.

- Identify any subsurface drainage and alterations that would need to be completed if a reservoir were to be constructed there.

- Select a location with clayey subsoil to allow for the construction of the reservoir from on-site water-holding materials. Importing clay from another site or using synthetic liners substantially adds to the project cost.

It is difficult to find a site that incorporates all of these requirements. Options and decisions will be influenced by cost and operational preferences. The goal is to choose the site that best incorporates all of the factors.

Site Investigation

A site investigation of the soil is extremely important to determine suitability for the construction of a water reservoir. Conduct an initial site assessment by checking existing soil information on available soil maps followed by a more detailed site investigation under the direction of an experienced soil professional. These professionals are usually found working in engineering consulting firms.

Required Soil Texture

Adequate clay content will minimize or eliminate water seepage through the bottom and sides of the reservoir. The ideal site would have subsoil with a minimum of 15% clay content, uniformly distributed throughout, to a depth greater than that of the proposed excavation.

With suitable clay content, the portion of the reservoir below the normal ground grade will retain water. Also, the excavated clay is used to construct the berms for the portion of the reservoir storage above the normal ground grade. Otherwise, clay needs to be found at another site to line the reservoir.

Investigation Process

A good site investigation includes taking test holes using backhoes, drills, augers or specialized boring machines so the soil samples can be assessed for their clay content, suitability for water retention, lining material and berm construction. Record the test holes locations, including GPS coordinates, along with descriptions of the soil profiles (called a drilling log).

Drilling for clay.

More advanced technology is also available to help with soil investigations. An electromagnetic (EM) survey is a technique which, without any excavation, can indicate where clay may be concentrated on the farm. The EM survey is a very useful tool for effectively targeting where to start the site investigation (excavations) in order to maximize knowledge of clay locations and minimize costs.

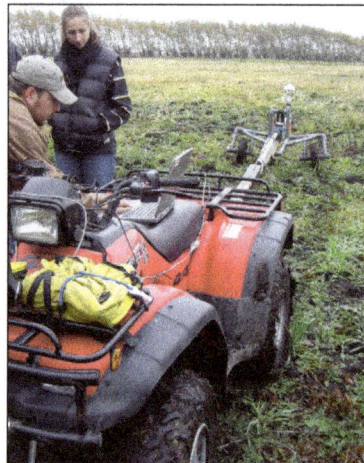
EM50 clay investigation.

It is critical to determine if there are any pipelines or utilities present that need to be avoided before beginning any excavations.

Implications of Findings

If the clay content is not suitable or there is not enough clay to line the reservoir and construct the berms, the site location for the reservoir may need to be moved, or another site may need to be investigated as a potential source of clay known as a "borrow pit".

If a reservoir needs to be lined, a 1 m layer of compacted clay is recommended. Adjust the reservoir depth and surface size depending on the quality and quantity of clay soil found.

Example of sufficient clay for lining a reservoir.

The test holes will also identify if groundwater may be an issue during the construction of berms, or when excavating the below-grade portion of the reservoir.

Design

When choosing a reservoir designer, select someone who is experienced and trained in berm, dyke or dam design. A professional engineer should do the design if the depth of water in the reservoir above normal ground grade is:

- More than 3 m. Reservoirs above grade pose a potential risk if the reservoir breaks and water is suddenly released.

- More than 1.5 m and the reservoir is close to buildings, infrastructure (roads, railway tracks, utilities) or places where people work, live or play. Consider that reservoir failure may be a risk to any infrastructure within a radius of 3x the reservoir length.

A good design and plan are important for the long-term successful operation of a water reservoir. The reservoir design should take the following into account:

- Selection of the reservoir depth and the berm height for.

- Water volume requirements.

- A favourable earth balance of soil excavated versus soil required for berm construction - otherwise, excess soil will need to be dealt with and any material that must be double-handled or hauled away adds to the cost to the project.

- Allowance for settlement after construction.

- The berm height allowance for freeboard needed to avoid overtopping by wave action or large rainfall (freeboard is the additional height above normal water level).

- The adequate top width for the proposed uses on the berm.

- Minimum width is usually 3m.

- A larger top width may be needed for the placement of pumps, maintenance access or other activities.

- The appropriate and safe interior and exterior slope selection (steeper slopes will use less land, decreasing cost, but are more prone to erosion and failure requiring increased maintenance costs).

- A key trench (?1 m deep, 3 m wide) is dug first into the subsoil along the centre line of the berm. This is repacked with compacted fill and serves to anchor the berm and prevent seepage below the berm.

- Pipes for filling and emptying the reservoir are placed over top of the berms. If the pipes need to be buried into the berms, place them well above the high-water level of the reservoir, otherwise they create a potential seepage path. Pipes through the berm and below the high water mark are possible but should be designed by a professional engineer.

Pipes for filling and emptying the reservoir are placed over
top of the berm (rather than through the berm).

Design variables for a combination storage type reservoir include:

- A reservoir depth that is 3-6 m below normal ground level (excavated depth).

- A berm height that is 4-6 m above normal ground level (>3 m should be designed by a professional engineer).

- An interior slope inclination from 6:1 (for lower maintenance) to 3:1 (may require riprap or other face protection to resist wave action, especially in larger water surfaces).

- Gentle slopes reduce erosion potential due to wave action and allow for better trafficability during construction.

- Wave action potential increases as the surface area of reservoir gets bigger.

- An exterior slope from 4:1 for easier slope maintenance to 2.5:1 for minimum land usage.

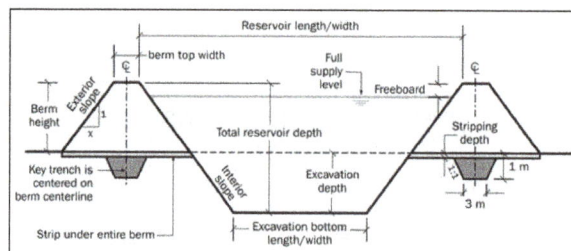

Design drawing for an above and below grade reservoir.

Example of an above and below grade reservoir.

Construction

Even the best design is only as good as the construction practices used when building the water reservoir. Request quotes from at least three contractors, view their work and get referrals. Consider the contractors experience, professional qualifications and specialized equipment. Place value on companies who have an engineering technician who supervises/inspects the construction. The following are key factors in the successful construction of a water reservoir.

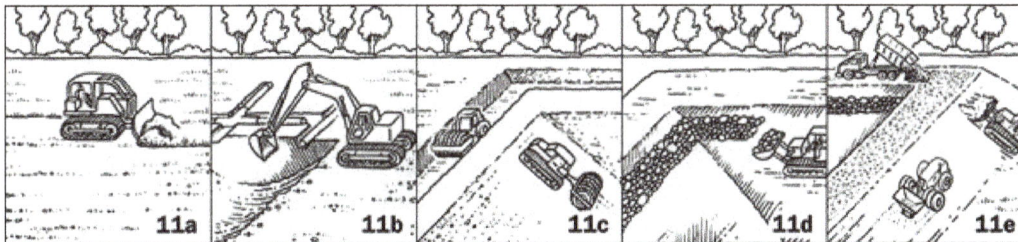
Components of construction process: Remove topsoil; Reroute tile drainage; Compact clay in key trench and on interior slope; Place rock rip rap at high water mark; Spread topsoil on banks and seed, place gravel on top of berm.

Equipment

Using proper equipment is important as large amounts of soil need to be excavated, moved, layered and properly compacted. Select a contractor who demonstrates experience and expertise in this type of reservoir construction and can implement the design, especially the berms.

Compaction

Starting with the key trench, place clay soil in layers 150-300 mm thick. Individually compact each layer. The clay must have the optimum moisture; this is confirmed by in-situ or laboratory tests done by the contractor, technician or engineer. The objective is to compact and smear the soil structure to create a relatively impervious soil to avoid any potential for seepage.

Compaction is achieved with specialized equipment, such as a "sheepsfoot" or "footed drum". Simply tracking with a bulldozer is not efficient and does not provide enough compaction. As the soil layers are built up and compressed, testing equipment can measure the density to ensure that compaction achieves the desired strength according to the design. Usually, 10 or more passes over the entire surface of each layer with the mechanical compactors, when the soil is at optimum moisture, will achieve the desired soil density.

Compacting soil using a sheepsfoot roller.

Erosion Control

The need for erosion protection at the waterline is not always clear. Flatter slopes do not always need protection with rock riprap. Protection is needed when the reservoir is kept full for most of the year. Since the main purpose is to provide a water supply for irrigation, the reservoir may not be full for long stretches of time, which could reduce the need for additional erosion protection. If rock riprap is used, it should always be underlain with a geotextile filter cloth. The filter cloth keeps the soil in place, and the rock riprap keeps the filter cloth in place while also providing protection from ice and waves.

Wave action erosion at the high water level.

Rock riprap used to prevent erosion at the high water level.

Groundwater

If groundwater is a problem near the berms, it should have been identified by the soil tests and provided for in the design. A perimeter drainpipe is often used to correct this situation.

Seeding

Once the berms are completed, top-dress the top and side slopes with a layer of topsoil and seed with a grass mixture within 24 hours, before the soil dries out. Choose a seed mixture that is drought tolerant and does not require mowing such as tall or creeping red fescue. Straw mulch, lightly spread, may help with early seedling establishment.

An above and below grade reservoir featuring a seeded bank.

Traffic

The top of the berm is often graveled to provide an all-weather access road.

Inspection and Maintenance

Inspect the reservoir on a regular schedule. Inspection in the spring (full) and fall (low) is a minimum. More frequent inspections should be done if there is any indication of damage or seepage.

Develop an inspection checklist and record of maintenance activities that will provide a historical record for review. A checklist will also demonstrate due diligence on the part of the operator. Timely maintenance, based on regular inspections, can address concerns in a cost-effective manner, before they develop into serious problems.

Failure and collapse of the interior bank of the reservoir.

- Inspection and Maintenance Checklist.

- Cracking or settling of the berms.

- Wet or soggy conditions at the berm toe.

- Stability of interior and exterior side slopes.

- Excessive erosion or sedimentation in or near the reservoir.

- Woody vegetation in or on the berms.

- Animal holes.

- Obstructions of the inlet or outlet devices by trash and debris.

- Deterioration of irrigation intake or pipes.

Soil Cracking

Soil cracking is often the result of the soil settling due to inadequate compaction or foundation compression. Soil cracking parallel to the crest of the berm reduces soil strength and the slope factor of safety, which could result in a slope failure depending on site-specific conditions. Soil cracking perpendicular to the crest of the berm may allow for seepage along the crack, again resulting in slope failure.

It is not easy to repair cracks. Although filling in the crack with clay helps to prevent water from entering or traveling along the crack, the effectiveness is limited. The water level in the reservoir may have to be reduced and/or the berm may need to be rebuilt.

Seepage

Water seepage reduces the safety factor of the slope and may cause slope failure. Seepage areas are identified by the presence of standing water, continuous wet conditions, water-loving vegetation and sometimes soil slumping.

Measures to lessen seepage problems post-construction are generally costly and are best prevented with proper initial design and construction techniques.

Erosion

Erosion can threaten the integrity of the berm. Seeding exposed slopes with a low-maintenance grass mixture is a normal part of construction to help prevent erosion before it begins. Repair eroded areas promptly as a normal part of maintenance, before they develop into a serious problem.

An example of a serious erosion problem because the bank was not re-seeded.

Organic Material

Sediment accumulation in the reservoir will depend on the type of material, the side slopes and the deposition of organic material such as leaves. Organic material entering the reservoir will build up over time and, as it breaks down, can impair the quality of the water. Eventually, if enough organics get into a reservoir, a cleanout may be required.

Unwanted Vegetation and Animals

Do not allow woody vegetation to grow on berms as, over time, roots can weaken the berm.

Mowing

Mow exterior slopes regularly unless a short grass has been seeded.

Safety

Each reservoir site location is unique and should be assessed for safety considerations. Remember that ponds are intriguing structures to children. Equip the site with the basic signage warnings and rescue equipment, such as a rope, a life buoy, a flotation device, etc. Consider installing a safety fence around the entire reservoir Check with the local municipality about reservoir safety requirements (fencing, etc.) and by-laws.

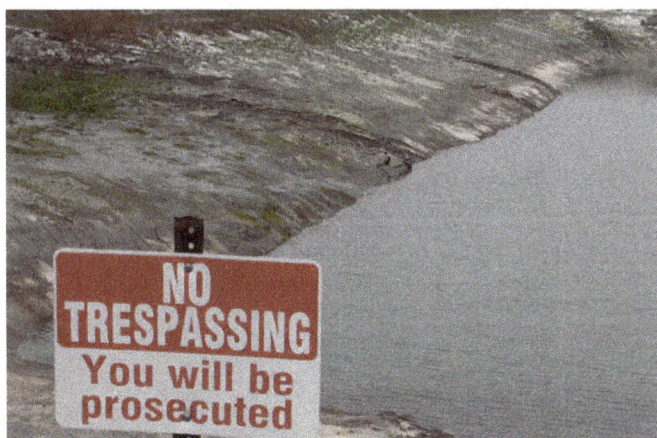

No trespassing sign posted at a reservoir.

Safety fencing installed around the entire reservoir.

Choice and Selection of Irrigation Methods

Compatibility of the Irrigation System

The irrigation system for a field or a farm must be compatible with the other existing farm operations, such as land preparation, cultivation, and harvest.

- Level of Mechanization,

- Size of Fields,

- Cultivation,

- Pest Control.

The use of the large mechanized equipment requires longer and wider fields. The irrigation systems must not interfere with these operations and may need to be portable or function primarily outside the crop boundaries (i.e. surface irrigation systems). Smaller equipment or animal-powered cultivating equipment is more suitable for small fields and more permanent irrigation facilities.

Topographical Characteristics of Area

Topography is a major factor affecting irrigation, particularly surface irrigation. Of general concern are the location and elevation of the water supply relative to the field boundaries, the area and configuration of the fields, and access by roads, utility lines (gas, electricity, water, etc.), and migrating herds whether wild or domestic.

Field slope and its uniformity are two of the most important topographical factors. Surface systems, for instance, require uniform grades in the 0-5 percent range.

Restrictions on irrigation system selection due to topography include:

- Groundwater levels.

- The location and relative elevation of the water source.

- Field boundaries.

- Acreage in each field.

- The location of roads.

- Power and water lines and other obstructions.

- The shape and slope of the field.

Economics and Cost of the Irrigation Method

The type of irrigation system selected is an important economic decision. Some types of pressurized systems have high capital and operating costs but may utilize minimal labour and conserve water. Their use tends toward high value cropping patterns. Other systems are relatively less expensive to construct and operate but have high labour requirements. Some systems are limited by the type of soil or the topography found on a field. The costs of maintenance and expected life of the rehabilitation along with an array of annual costs like energy, water, depreciation, land preparation, maintenance, labour and taxes should be included in the selection of an irrigation system.

Main costs include:

- Energy

- Water

- Land Preparation

- Maintenance

- Labor

- Taxes.

Soils

The soil's moisture-holding capacity, intake rate and depth are the principal criteria affecting the type of system selected. Sandy soils typically have high intake rates and low soil moisture storage capacities and may require an entirely different irrigation strategy than the deep clay soil with low infiltration rates but high moisture-storage capacities. Sandy soil requires more frequent, smaller applications of water whereas clay soils can be irrigated less frequently and to a larger depth. Other important soil properties influence the type of irrigation system to use.

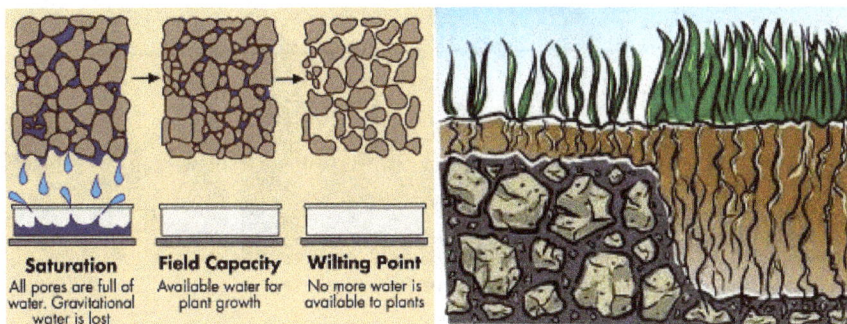

Saturation All pores are full of water. Gravitational water is lost

Field Capacity Available water for plant growth

Wilting Point No more water is available to plants

The physical, biological and chemical interactions of soil and water influence the hydraulic characteristics and filth. The mix of silt in a soil influences crusting and erodibility and should be considered in each design. The soil influences crusting and erodibility and should be considered in each design. The distribution of soils may vary widely over a field and may be an important limitation on some methods of applying irrigation water.

The soil type usually defines:

- Soil moisture-holding capacity.

- The intake rate.

- Effective soil depth.

Water Supply

The quality and quantity of the source of water can have a significant impact on the irrigation practices. Crop water demands are continuous during the growing season. The soil moisture reservoir transforms this continuous demand into a periodic one which the irrigation system can service. A water supply with a relatively small discharge is best utilized in an irrigation system which incorporates frequent applications. The depths applied per irrigation would tend to be smaller under these systems than under systems having a large discharge which is available less frequently. The quality of water affects decisions similarly. Salinity is generally the most significant problem but other elements like boron or selenium can be important. A poor quality water supply must be utilized more frequently and in larger amounts than one of good quality.

Crops to be Irrigated

The yields of many crops may be as much affected by how water is applied as the quantity delivered. Irrigation systems create different environmental conditions such as humidity, temperature, and soil aeration. They affect the plant differently by wetting different parts of the plant thereby introducing various undesirable consequences like leaf burn, fruit spotting and deformation, crown rot, etc. Rice, on the other hand, thrives under ponded conditions.

Some crops have high economic value and allow the application of more capital-intensive practices, these are called "cash crops" or Cash crop farming. Deep-rooted crops are more amenable to low-frequency, high-application rate systems than shallow-rooted crops.

Cash Crop Water Requirement.

Crop characteristics that influence the choice of irrigation system are:

- The tolerance of the crop during germination, development and maturation to soil salinity, aeration, and various substances, such as boron.

- The magnitude and temporal distribution of water needs for maximum production.

- The economic value of the crop.

Social Influences on the Selection of Irrigation Method

Beyond the confines of the individual field, irrigation is a community enterprise. Individuals, groups of individuals, and often the state must join together to construct, operate and maintain the irrigation system as a whole. Within a typical irrigation system there are three levels of community organization. There is the individual or small informal group of individuals participating in the system at the field and tertiary level of conveyance and distribution. There are the farmer collectives which form in structures as simple as informal organizations or as complex as irrigation districts. These assume, in addition to operation and maintenance, responsibility for allocation and conflict resolution. And then there is the state organization responsible for the water distribution and use at the project level.

Irrigation system designers should be aware that perhaps the most important goal of the irrigation community at all levels is the assurance of equity among its members. Thus the operation, if not always the structure, of the irrigation system will tend to mirror the community view of sharing and allocation.

Irrigation often means a technological intervention in the agricultural system even if irrigation has been practiced locally for generations. New technologies mean new operation and maintenance practices. If the community is not sufficiently adaptable to change, some irrigation systems will not succeed.

External Influences

Conditions outside the sphere of agriculture affect and even dictate the type of system selected. For example, national policies regarding foreign exchange, strengthening specific sectors of the local economy, or sufficiency in particular industries may lead to specific irrigation systems being utilized. Key components in the manufacture or importation of system elements may not be available or cannot be efficiently serviced. Since many irrigation projects are financed by outside donors and lenders, specific system configurations may be precluded because of international policies and attitudes.

Crop Water Requirement in Irrigation

It is defined as, "The quantity of water required by a crop in a given period of time for normal growth under field conditions." It includes evaporation and other unavoidable wastes. Usually water requirement for crop is expressed in water depth per unit area.

Irrigation Water Need = Crop water need — available rain fall.

The first thing you need to consider when planning your garden is what growing zone you live in. This is based on both the temperature range of your climate and the amount of precipitation. Take a close look at the area in which you are going to plant your garden. If the ground tends to be very moist, choose plants that can tolerate constantly wet soil, and even standing water.

If you live in an area that suffers from frequent droughts, however, select plants that can tolerate going long periods without water, especially in light of the frequent watering restrictions imposed on such areas. If you are lucky enough to live in an area that has a balanced climate, you have a wider range of choices for your plants.

Low Water Requirement Plants

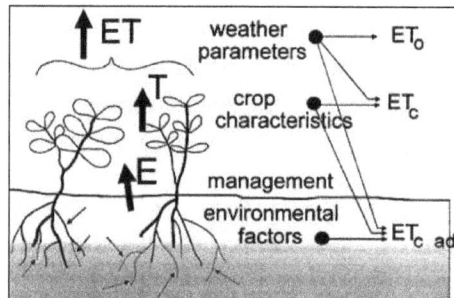

Plants that require low levels of water are often called drought tolerant. Drought-tolerant plants can thrive in hot, dry conditions with very little water. They include both perennials and annuals. Most drought-tolerant plants only have to be hand-watered when they are planted and while they are establishing themselves. After that, they can be left to the natural cycle of the elements. Popular drought tolerant trees include the red cedar. live oak, crape myrtle, and the windmill and saw palmetto palm trees. All citrus trees are also drought tolerant. Many homeowners in areas prone to drought, such as parts of the southern United States, use shrubs and ground covering vines as part of their landscaping. These include Texas sage, orange jasmine and Chinese fountain grass. There are not many perennial drought-tolerant plants, but amaryllis is one that is very popular, along with the African iris. Popular drought-tolerant annuals include marigold, cosmos and the Dahlberg daisy.

Mid-level Water Requirement Crops

Most plants land in this range when it comes to water requirements. These plants do not need to be watered every day, but they need to be watered when the soil has been dry for over a week or two. Sometimes these plants are classified as plants lying in the "occasional water zone". These include popular plants such as geraniums, most roses, wisteria, clematis and other vine plants, sunflowers, spring flowering bulbs, and most flowering perennial shrubs. Note that flowering annuals planted in containers will need watering at least once or twice a week, while annuals planted in the ground will need watering less often.

High Water Requirement Plants

Some plants require large amounts of water. These plants typically grow in marshy areas or bogs, or along the banks of rivers, streams and lakes. The soil for these plants should always be kept moist. Standing water is not a concern for these plants, so you don't have to worry about root rot. Perennials are especially good for wet areas because they don't have to be replanted year after year, which can be difficult in marshy areas. Popular perennials for wet soil include iris plants, cannas, bee balms, ferns, and bog salvia. Aquatic mint is a pleasant ground cover that likes wet soil. The red osier dogwood does very well in wet conditions. Most annual flowering plants also do well in constantly moist soil.

Water Requirement of Different Crops

Amount of water required by a crop in its whole production period is called water requirement. The amount of water taken by crops vary considerably. What crops use more water and which ones less.

Crop	Water Requirement (mm)
Rice	900-2500
Wheat	450-650
Sorghum	450-650
Maize	500-800
Sugarcane	1500-2500
Groundnut	500-700
Cotton	700-1300
Soybean	450-700
Tobacco	400-600
Tomato	600-800
Potato	500-700
Onion	350-550
Chillies	500
Sunflower	350-500
Castor	500
Bean	300-500
Cabbage	380-500
Pea	350-500
Banana	1200-2200
Citrus	900-1200
Pineapple	700-1000
Gingelly	350-400
Ragi	400-450
Grape	500-1200

Irrigation Crop Water Requirement

This case study shows how to calculate the total water requirement for a command area (irrigation blocks) under various crops, soil textures and conveyance loss conditions. In order to evaluate the required irrigation gift for the entire command area a simple water balance has to be set-up. The total water demand for each irrigation block and the crops in each block are calculated by summing the following components:

- Infiltration (percolation loss) through the soil (I).

- Seepage (conveyance loss) through the channel (S).

- Maximum evapo-transpiration of the crop (ETm).

In this exercise, the irrigation water requirement is calculated for a 10-day period during the harvest stage.

Evaluation of Percolation Loss (I)

The command area is divided in irrigation blocks. First, these irrigation blocks are crossed with the soil texture map to determine the area of each soil texture class in each block. Percolation losses differ per soil texture class so a table with the following percolation data is created:

Texture	Percolation loss (mm/day)
Clay	4
Loam	12
Sandy clay	14
Clay loam	7

The percolation table is joined with the cross table to get the percolation for each soil texture class in each block. The amount of water loss for each soil texture class per block is calculated with a tabcalc statement. In order to get the total percolation loss per block the results of the previous operation are aggregated.

Evaluation of Conveyance Loss (S)

Conveyance losses are calculated in about the same way as the percolation losses. First, the map with the irrigation blocks is crossed with the channel distribution map. The conveyance loss per meter channel length differs per channel type and is 0.2 m^3 per day for clay channels and 0.01 m^3 per day for concrete channels. A new table indicating water loss per channel type is created and joined to the cross table. The amount of water loss for each type of channel per block is calculated with a simple tabcalc formula. Finally the results are aggregated to evaluate the total conveyance loss per irrigation block.

Evaluation of Maximum Evapo-transpiration (ETm)

Crop water requirements are normally expressed by the rate of evapotranspiration (ET). The evaporative demand can be expressed as the reference crop evapotranspiration (ETo) which predicts the effect of climate on the level of crop evapotranspiration. In this case study the ETo is 8 mm/day. Empirically-determined crop coefficients (kc) can be used to relate ETo to maximum crop evapotranspiration (ETm) when water supply fully meets the water requirement of the crop. The value of kc varies with crop and development stage.

For a given climate, crop and crop development stage, the maximum evapotranspiration (ETm) in mm/day of the period considered is:

$$ET_m = k_c * ET_o$$

Maximum evapo-transpiration refers to conditions when water is adequate for unrestricted growth and development under optimum agronomic and irrigation management. Maximum evapotranspiration is calculated in this case study by crossing the irrigation block map with the map that shows the different crop types in the command area, joining the cross table with the kc table and by applying the maximum evapotranspiration formula with a tabcalc statement.

Water Balance Calculation (S+I+ETm)

The required irrigation gift for the entire command area is equal to the sum of water losses due to infiltration through the soil (I), seepage through the channel (S) and maximum evapotranspiration (ETm) for each block. The total amount of water requirement in harvest period for each block is reclassified in irrigation classes using the following table:

Upper boundary	Irrigation class
4000	1
6000	2
8000	3
10000	4
12000	5
14000	6

Finally, you will create a script to automate the calculation procedure. With the script, you can easily calculate the water requirements for other growing stages.

Irrigation Scheduling

The purpose of irrigation scheduling is to determine the exact amount of water to apply to the field and the exact timing for application. The amount of water applied is determined by using a criterion to determine irrigation need and a strategy to prescribe how much water to apply in any situation.

Irrigation Criteria and Irrigation Scheduling

Irrigation criteria are the indicators used to determine the need for irrigation. The most common irrigation criteria are soil moisture content and soil moisture tension. The less common types are irrigation scheduling to maximize yield and irrigation scheduling to maximize net return. The final decision depends on the irrigation criterion, strategy and goal. Irrigators need to define a goal and establish an irrigation criterion and strategy.

To illustrate irrigation scheduling, consider a farmer whose goal is to maximize yield. Soil moisture content is the irrigation criterion. Different levels of soil moisture trigger irrigation. For example, when soil water content drops below 70 percent of the total available soil moisture, irrigation should start.

Soil moisture content to trigger irrigation depends on the irrigator's goal and strategy. In this case, the goal is to maximize yield. Therefore, the irrigator will try to keep the soil moisture content above a critical level. If soil moisture level falls below this level, the yield may be lower than the maximum potential yield. Thus, irrigation is applied whenever the soil water content level reaches the critical level.

How much water to apply depends on the irrigator's strategy. For example, the irrigator can replenish the soil moisture to field. capacity or apply less. If no rain is expected and the irrigator wishes to stretch the time between irrigations, it is advantageous to refill the soil profile to field capacity. If rain is expected, it may be wise not to fill the soil profile to field capacity, but leave some room for rain.

If the irrigator's goal is to maximize net return, an economic irrigation criterion is needed, such as net return. This is the income from the crop less the expenses associated with irrigation.

The importance of irrigation scheduling is that it enables the irrigator to apply the exact amount of water to achieve the goal. This increases irrigation efficiency. A critical element is accurate measurement of the volume of water applied or the depth of application. A farmer cannot manage water to maximum efficiency without knowing how much was applied.

Also, uniform water distribution across the field is important to derive the maximum benefits from irrigation scheduling and management. Accurate water application prevents over- or underirrigation.

Overirrigation wastes water, energy and labor; leaches expensive nutrients below the root zone, out of reach of plants; and reduces soil aeration, and thus crop yields. Underirrigation stresses the plant and causes yield reduction.

Advantages of Irrigation Scheduling

Irrigation scheduling offers several advantages:

1. It enables the farmer to schedule water rotation among the various fields to minimize crop water stress and maximize yields.

2. It reduces the farmer's cost of water and labor through fewer irrigations, thereby making maximum use of soil moisture storage.

3. It lowers fertilizer costs by holding surface runoff and deep percolation (leaching) to a minimum.

4. It increases net returns by increasing crop yields and crop quality.

5. It minimizes water-logging problems by reducing the drainage requirements.

6. It assists in controlling root zone salinity problems through controlled leaching.

7. It results in additional returns by using the "saved" water to irrigate non-cash crops that otherwise would not be irrigated during water-short periods.

Research in Nebraska, where most water is pumped, shows that irrigation scheduling provides an average 35 percent savings in water and energy. In fuel costs alone, this is a per-season savings of about 550 kwh per acre for a center pivot sprinkler or about 200 kwh per acre for a gated pipe.

Irrigation Scheduling Methods

All irrigation scheduling methods consist of an irrigation criterion that triggers irrigation and an irrigation strategy that determines how much water to apply. Irrigation scheduling methods differ by the irrigation criterion or by the method used to estimate or measure this criterion. A common and widely used irrigation criterion is soil moisture status.

Construction of an Evaporation Pan for Irrigation Scheduling

Plants suck up water from the soil and lose the majority of this water to the air through their leaves by a process known as evapotranspiration. Ideally we would like to measure plant water use so that irrigations can be accurately scheduled, however, the technology is complex and currently not suited to practical irrigation management. Alternatively, plant water use can be estimated from the measurements of evaporation from an open surface of water. From evaporation readings irrigations can be timed so that the soil is at the right moisture level and irrigations run long enough to replace the water the plants have used.

There are many types of evaporation pans used by farmers. However, the universal pan is the United States Weather Bureau Class A pan evaporimeter. It is important to use the same dimensions as this universal pan, mainly because the effect of wind and temperature on evaporation will vary with the surface area and the depth of water in the pan. Evaporation and irrigation replacements cannot be compared between sites if non standard pans are used.

Evaporimeter.

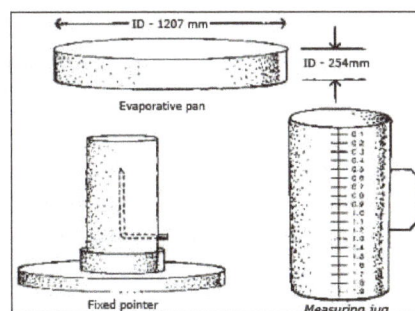

Components of the class a pan evaporimiter.

Construction

There are three parts to an evaporimeter. All parts can be made very cheaply with common materials. Alternatively a complete unit can be purchased at considerably greater cost. The following is a description of how to construct the three components of the evaporimeter.

Evaporation Pan

The evaporation pan must be made to the standard specifications of an internal diameter of 1207 mm and height of 254 mm using 20 gauge galvanised iron. The standard material is galvanised iron as alternatives will have different thermal and reflectance properties, therefore altering the evaporation rate. It is best to have the pan made by either a galvanised tank manufacturer or an engineering firm. Before the pan is sited in the field it should be checked for leaks.

Fixed Pointer

The fixed pointer that sits inside the pan can be made from standard irrigation fittings and a piece of stainless steel rod. There are three parts to the fixed pointer:

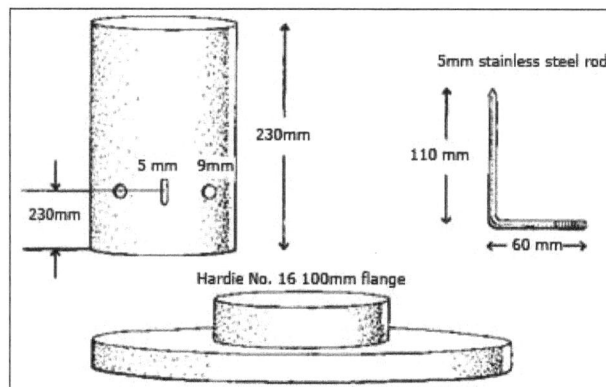

Components of the fixed pointer.

- The base, a 100 mm PVC flange.

- The pointer support, a 230 mm long piece of 100 mm PVC pipe. Four equally spaced 9 mm holes 70 mm from the base are drilled to allow the water height around the fixed pointer to quickly adjust to the water height in the pan. A single 15 mm long 5 mm wide elongated hole is also drilled 70 mm from the base of the PVC pipe.

- The pointer, a 170 mm long piece of 5 mm stainless steel rod bent at a right angle 60 mm from one end. From the shorter end a thread is tapped for about 15 mm and a point is ground on the other end of the rod.

After fitting the PVC pipe into the flange, the stainless steel rod is inserted into the elongated hole with nuts located on the inside and outside of the PVC pipe. To initially set the stainless steel rod in the correct position, the fixed pointer is placed in the pan and the pan is filled with water to a depth of 190 mm. The rod is then slid up or down in the 5 mm elongated hole so that the point of the rod just breaks the surface of the water.

Measuring Cylinder

To measure evaporation the pan must be refilled with a known volume of water. The surface area of the pan is 1.14 square metres, so for every mm of evaporation 1.14 litres of water must be added to the pan. A transparent plastic 2 litre measuring jug with vertical sides is an excellent measuring cylinder if it is scaled properly. It is important that the jug actually holds more water than 2 litres so the sides of the jug must extend past the 2 litre mark. The jug is filled with 2.28 litres of water and the water level marked. This can conveniently be done by weighing the jug and adding 2.28 kilograms of water. For most jugs this will just about overflow, which is perfect.

A jug of water filled to the marker will be equivalent to 2 mm of evaporation. To scale the jug when less than 2 mm of water is required to fill the pan, the distance from the top marker to the bottom of the jug is measured and divided by 20. The numbers 0 to 2.0 in increments of 0.1 are then written with a permanent marking pen from the top marker to the bottom of the jug. These numbers are equivalent to the same number of mm of evaporation from the pan.

Measurement

With evaporation the water level in the pan will fall. To measure the amount of evaporation, water is added to the pan with the measuring jug filled to the top mark. Water is added until the pointer just breaks the surface of the water. The PVC pipe supporting the pointer will help by reducing wave motion. It is important to keep track of the number of jugs used to refill the pan and the reading on the last jug when the pan water level is just broken by the pointer. The total amount of water added equals the amount of evaporation.

It is also essential to measure rainfall in conjunction with evaporation. Both measurements enable evaporation to be calculated on rainy days. After heavy rain the pan may have to be emptied to bring the water level down to the pointer. After rainfall on a hot summer's day, less water may have to be removed than actually fell as rain. For example, after a 25 mm rainfall there might only be 12 mm of water removed from the pan with the measuring jug to bring the water level back to the pointer. The difference between the rainfall (25 mm) and the water removed from the pan (12 mm) is the evaporation. In this example it is 13 mm.

If the rain does not fill the pan above the pointer, the rainfall must still be added onto the measured evaporation to give the actual evaporation. For example, if there was 7 mm of rainfall and 6 mm of water was added to the pan with the measuring jug then the evaporation would be 13 mm.

Evaporation measurements should be routinely done every day at 9.00 am and clearly recorded. If measurements are not done routinely then the volume of water in the pan will decrease and take less time to heat up during the day and cool at night. This will induce an error which will become greater as the volume of water in the pan decreases. Evaporation measurements are very simple and take less than 5 minutes.

Siting

The pan should be placed on a flat wooden platform about 150 mm above the ground surface. A pallet is perfect. To avoid animals and birds interfering with the water in the pan we strongly advise covering the pan with chicken wire. This is not a USWB standard but is recommended for all agricultural situations. The error from covering the pan with bird wire will be small, much less than that introduced by animal interference.

To construct a simple chicken wire cage over the pan, bend a piece of 6 mm steel rod around the outside of the pan, welding the ends together to form a hoop. Chicken wire is then loosely tied to the perimeter of the hoop. This cage can then be slipped over the pan.

The area around the pan must be grassed and free from weeds, bushes and trees so that no shading can occur. The grass around the pan must be mown regularly and kept green. A nearby water supply is advised for refilling the pan and watering the grass around the pan.

Maintenance

The pan should be cleaned at least once a year. After cleaning and every month an algaecide should be added at the same rate as for swimming pools. If an algaecide is not added then the pan should be emptied and cleaned monthly.

If the maintenance or siting of the pan or the construction of the pan is not standard, then there will be a different relationship between evaporation and plant water use, but this difference is not critical if the differences are small and consistent. Variations from month to month like algae growth, unmown grass or reading at very irregular intervals are far worse than a consistent difference. For example we suggest that if the grass cannot be kept green and mown then the area around the pan should be kept bare with a herbicide. It should be recognised here that non-standard evaporation pans can still be very useful for irrigation scheduling. They must, however, be evaluated on their own. Published irrigation replacement factors will be less accurate than for a standard evaporation pan. We recommend that if a non standard evaporation pan is used for irrigation scheduling then soil moisture and plant performance must be monitored to determine appropriate irrigation replacements from evaporation.

References

- Irrigation-engineering-importance: omicsonline.org, Retrieved 2 March, 2019
- Planning-of-an-irrigation-project, irrigation-project, irrigation: biologydiscussion.com, Retrieved 12 June, 2019
- Types-canal-construction-machines, construction: theconstructor.org, Retrieved 6 February, 2019
- Assessment-of-charges-for-irrigation-water, irrigation: biologydiscussion.com, Retrieved 8 March, 2019
- Engineer: omafra.gov.on.ca, Retrieved 11 August, 2019
- Factors-affecting-choice-of-selection-of-irrigation-method: aboutcivil.org, Retrieved 14 January, 2019
- Water-requirements-of-crops: aboutcivil.org, Retrieved 20 May, 2019
- Crops: colostate.edu, Retrieved 7 July, 2019
- Construction-of-an-evaporation-pan-for-irrigation-scheduling, irritation, soil-and-water, farm-management, agriculture: vic.gov.au, Retrieved 17 April, 2019

Chapter 3

Irrigation: Operations, Processes and Systems

There are various processes which are a part of irrigation such as water conveyance, irrigation scheduling and designing open channels. Some of the systems which are used in irrigation are automatic plant irrigation system, automated rain barrel drip system and underground pipeline system. This chapter discusses in detail these processes and systems related to irrigation.

Water Conveyance and Distribution Canals

Conducting network in open irrigation systems is composed of the main canal, inter-farm, farm, and on-farm distribution canals (distributors) of different orders. The main canal delivers water from a river, reservoir, well etc. to inter-farm distributors which, in turn, deliver water to individual farms or crop-rotation sites; on-farm distributors deliver water to crop-rotation fields or irrigated sites. In some cases, the conducting network has no full set of canal structures.

Irrigation canals are arranged so that to provide the following at minimum construction and operation costs: water supply of required volume and in due time; highest efficiency of the canals (ratio of the flow rate at the canal tail to that at the canal head) and irrigated area use ration; effective operation of the canals and structures in those. Command of the main canal (excess of the water level in it above the water level in lower-order canals) over the respective irrigated area and higher-order canals over lower-order canals) is essential for the irrigation network operation ensuring gravity irrigation.

Inter-farm Canals

Inter-farm canals are the canals that distribute the water delivered to the main canal to all farms within the system.

In the open irrigation network, the conducting network is composed of the main canal, inter-farm, farm, and on-farm distributors of different orders. Water is supplied through the main canal from an irrigation source to the inter-farm distributors that deliver water to certain land users or crop-rotation lands; on-farm distributors deliver water to fields or irrigated sites. Canal routes are designed along the boundaries of farms, crop-rotation lands, fields, and so on in order to keep the integrity of the irrigated area.

The inter-farm canal is a branch of the main canal and distributes the water delivered through it among certain farms or crop-rotation sites. Here a branch means a largest canal outgoing from the main canal, according to the significance (in terms of discharge, serviced area, length) of which it might be considered as an extension of the main canal.

Canals-distributors represent smaller sections of the inter-farm and farm irrigation network. The routes of irrigation canals should pass along the boundaries of farms, crop-rotation sites, fields so that not to split the respective irrigated area, while the main canal – by the highest points of the irrigated area.

Irrigation canals are laid through an excavation, cut-and-fill, or on a hillside. To reduce seepage losses, the canals' bottoms and walls are compacted, covered with facing from concrete, (cast-in-situ reinforced or precast structural) reinforced concrete, and shields from clay and polymer films are applied.

On-farm Canals

On-farm canals are the ones that deliver water to each farm and, if a farm is big enough, to its separate large irrigated sites, without hampering the mechanization of farm operations and allowing treating of cultivated crops.

On-farm irrigation networks consist of on-farm and delivery ditches as well as temporary irrigation network.

Permanent canals distribute water over the farm's territory: they are designed on the maps to a scale of 1:10000 or 1:5000 with the contour intervals of 0.5 m, when the territory area, crop rotation pattern, irrigation regime, and method of irrigation of all crops are known.

Designing of the network begins from plotting farms' boundaries on the map and determining the total irrigation area and the area usable for irrigation. Then the routes of on-farm distributors are outlined according to the highest elevation of the relief with simultaneously setting crop rotation sites' boundaries.

Crop rotation sites' boundaries are marked so that they match higher-order canals, roads, boundaries of human settlements, natural boundaries and taking into account the hydrogeological conditions and the ease of use of the system. After farms' boundaries and approximate boundaries of crop rotation sites are set and on-farm distributors are routed, they proceed to laying out crop rotation fields and irrigated sites within each crop rotation process.

Temporary Irrigation Network

Temporary irrigation network means the network that carries out regulation within irrigated sites, composed of temporary, annually installed irrigation ditches, field (auxiliary) ditches and irrigation furrows and irrigation ditches, distribute water to fields and, when irrigating, transfer it into soil moisture of required extent. Irrigation and regulation network can also be composed of closed or mobile pipelines and mobile sprinkler plants, and of underground pipelines at subsoil irrigation.

The temporary irrigation ditches and field head ditches located on an irrigated site are cut and leveled by tractor-drawn universal ditchers as necessary.

Ploughing, harrowing, sowing, cultivation, extra nutrition, and harvesting on an irrigated site are performed by means of tractor units. Since the operational efficiency of tractor units at ploughing and other works considerably decreases at the run (pass) length of less than 400 m, it is advisable to take the area of the given site not less than 400 x 400 m².

The requirements applicable to the temporary irrigation network. The temporary irrigation network is composed of temporary irrigation ditches, field and furrow ditches which are used for water withdrawal from delivery ditches and supply it to irrigated irrigation ditches and furrows at surface irrigation and to sprinklers at sprinkling irrigation.

Temporary canals must meet the following requirements: pass design flow; not be washed away; have no adverse slopes; command the adjacent areas; be straight; parallel to each other; and, as far as possible, parallel to the sides of the irrigated site.

Temporary canals are arranged by two schemes: longitudinal and transversal.

With the longitudinal scheme, temporary irrigation ditches are laid along furrows or irrigation ditches (perpendicularly to the horizontals), and discharge furrows – across furrows or irrigation ditches (at an acute angle to the horizontals).

Water from the delivery ditch runs through a water (drainage) outlet to a temporary irrigation ditch, from which flows to a discharge furrow, and from the discharge furrow to irrigated furrows or irrigation ditches by means of siphons, pipes, and other devices.

To avoid erosion of temporary canals the longitudinal scheme is applied on sites with a slope gradient of not more than 0.004 at an irrigation ditch discharge of up to 80 l/s and 0.003 at a discharge of over 80 l/s, since they are eroded at steeper slopes.

The length of temporary irrigation ditches are taken equal to 400-800 m for the crops requiring inter-row cultivation and 400-1500 m for narrow-row crops and grasses. The distance between them with single-sided command is 70-200 m and over, while between the field head ditches it equals the length of the irrigated furrows or irrigation ditches.

When irrigating by long furrows, the temporary irrigation ditch is allowed to operate for up to two days.

With the transversal scheme, temporary irrigation ditches are trenched across irrigated furrows or irrigation ditches at a gradient of 0.001-0.002, and water is supplied to the irrigated furrows or irrigation ditches from temporary irrigation ditches. The transversal scheme of temporary irrigation ditches layout is applied on sites with a gradient of more than 0.004.

The distance between the temporary irrigation ditches equals the length of the irrigated furrows or irrigation ditches. The length of the temporary irrigation ditches is taken equal to 400-1000 m depending on the degree of leveling-off degree of the site, irrigation ditch grade, water discharge, soil permeability.

With the transversal scheme of irrigation ditches layout the works related to bedding up and leveling of the field head ditches are excluded, the performance factor of the temporary irrigation network on the irrigated site as well as land use ratio are improved.

With the transversal scheme, permanent canals with a grade of 0.0005 for a discharge of 200-300 l/s are laid instead of temporary irrigation ditches. Water from the permanent canals is supplied through water outlets to the sections of the single-side furrow ditch from which it is distributed among irrigated furrows. With canal length of 1200-1500 m and furrow length of 300-500 m, an irrigated site with an area of 36-75 ha forms, which meets the agricultural work mechanization requirements. The discharge of 200-300 l/s is controlled by a field irrigator, who applies water at a rate of 1 ha/h.

Preparation of the areas of irrigated sites for water application. It includes: one-time-only construction leveling of the site surface; periodic operating leveling and surfacing of the field; cutting and leveling of the temporary irrigation network (irrigation ditches, irrigated furrows, and check plots); installation of reinforcing bars on the irrigation network.

When constructing or reconstructing an irrigation network, fundamental leveling is carried out according to the design. Operating leveling is carried out periodically by using tractor-mounted buckrakes, levelers, long-span blade levelers, executing in summer or after harvesting, or in spring before sowing late crops.

Irrigation ditches are trenched simultaneously with the sowing process, and broad irrigation ditches before water application. Irrigation furrows are cut before water application.

Field head ditches and temporary irrigation ditches are cut: after sowing grain crops and grasses; before the first water application after irrigation furrows are cut; before each regular watering if ones (ditches) had partially been destroyed in the course of previous water application and cultivation processes.

Prior to water application, they prepare and install irrigation equipment: siphons; gates; dikes (closure dams); irrigation pipes and gates; weirs.

The field head ditches and temporary irrigation ditches are leveled after the last vegetative watering in every case, except for the fields under permanent grasses; before a cultivation process if the temporary irrigation ditches complicate the passage of tractors; after charging (off-season) irrigation and leaching of saline lands.

Water-diversion Network

Water-diversion network is composed of canals of different orders, among which are field drainstail ditches; land plot, on-farm, and inter-farm collectors. The highest-order section of the water-diversion network is the main collector that diverts drainage & escape waters to an intake conduit.

The water-diversion network on rice fields must perform the following functions: rapidly lower groundwater level to the level depth where it is possible to carry out mechanized operations to sow and harvest rice under saline soil conditions and mineralized groundwater; perform desalting drainage functions; provide soil leaching in the aeration zone; prevent repeated soil salinization on the fields of crop rotation with rice not covered by rice; and control redox processes in aeration zone soils. The peripheral canals of a water-diversion network must guard the areas adjacent to the rice field against flooding and salinization as well as protect the rice-growing site from inflow of groundwater and storm water to the territory of the rice growing system from outside.

Tailwater Network

Tailwater network serves for removing excessive surface water (formed at emptying canals, in emergency cases, at downpour, etc.) out of irrigated lands and is placed orienting by lower check points of the irrigated lands.

The tailwater network (in an irrigated zone) is the system of the ditches and facilities designed for intake and transit pass of mud stream, emergency, storm waters, as well as for diverting excessive water from canals and out of fields; it is a component of the irrigation system. In submountain areas, natural depressions (ravines, gullies) serve as tailwater networks; only in depressions with large catchment basins (sais) and frequent mud torrents, special discharge ditches (waste ditches/ escape ditches), anti-mudflow structures (dams, dikes, water outlet conduits, spillovers). Local discharge of melt water, storm water, and irrigation water from fields and canals inflows to natural depressions, where it partially evaporates, percolates deep into ground and partially comes to downstream irrigation networks. On ravine and delta lands that are bad in terms of ameliorative conditions, collector and drainage network acts as drainage network.

The tailwater network represents the system of ditches designed to remove excessive surface and drainage waters out of irrigated lands. The tailwater network consists of: interception drains meant for catchment and diversion of surface water; emergency water removal and tail escapes for emptying permanent irrigation canals; water outlets that collect and divert excessive surface water.

Interception (catchwater) drains are laid upstream irrigated lands, while emergency water removal and tail escapes as well as water outlets orienting by lower check points of the topography. The routes of waste ditches are advisable to be laid on water users' boundaries and link with the location of the distribution network. Escape structures are also arranged in particularly dangerous places along the main canal and main distributors; to empty those, they use flood gates which are arranged, as appropriate, opposite natural thalwegs. If such facilities are absent, they make waste ditches that connect the flood gate with an intake conduit.

Water-collecting and Escape Network

The water-collecting & escape network of ditches is designed to organize catchment and diversion the following waters from the area of the irrigation system: surface flow (storm and melt water); water from distributors and irrigation ditches at process water disposal, emptying, and disasters; waster water from fields at surface irrigation and sprinkling irrigation.

Water-collecting & escape network is supposed to provide timely water diversion to an intake without violating the operation mode of irrigation system facilities and flooding irrigated lands, have minimum length and number of intersections with irrigation and collector & drainage network, communications. It is designed to be placed along the boundaries of irrigated sites, crop-rotation fields, as a rule, on subdued topographical features (backlands).

Water-collecting & escape network consists of:

- Field catchment areas that collect excessive water from fields;

- Conducting wasteway channels that divert water from field catchment areas, emergency water removal and tail escape structures on canals;

- Enclosing canals that protect irrigated lands against inflow of surface water from upstream areas.

Field catchment areas represent generally furrows cut at the low field edge which divert out the water collected to a lower-order waste ditch which is usually overlapped with a ditch of a field road.

A high-order escape network may consist of earth (unlined) canals, conduits. To divert water from the pipe network, they lay discharge conduits that run from discharge wells on irrigation pipelines to the closest waste ditch at the field boundary.

The highest surface flow rate of 10% water availability in the territory of the irrigated site or surface flow at irrigation is taken as the design flow rate in the canal of the water-collecting & escape network depending on the location and order of the canal.

The design flow rate of the emergency discharge from the main canal is equal to 0.5Qmax in the emergency discharge canal.

The design flow rate in water-collecting & escape ditches is taken as equal to not more than 30% of the total maximum discharge flows from simultaneously operating irrigation canals that discharge water to those. The design water flow of tail escapes for earth canals is taken as equal to 0.25-0.5 of the maximum discharge flow in permanent irrigation canal at the end section. In some cases, water-collecting & escape/waste ditches are checked in terms of passing melt or storm water from the areas they service.

Ditch parameters are set according to the maximum flow rate at uniform operation regime. Passage of design maximum flow rates through the canal is provided for a water level 15-20 cm below the ground surface. In rice-growing irrigation systems, they necessarily design a water-diversion (escape) network for which maximum and minimum flow rates are set.

The water level in the water-diversion canal of all orders is designed for 0.5 m lower the surface level of the lowest check adjacent to it.

Purpose and location of the water-collecting & escape network. At downpour, irrigation with discharge, emptying canals after irrigation, and water seepage from canals, surface runoffs collect in the irrigation systems located in depressions. They feed groundwater and may serve as the malarial mosquito breeding site. To remove excessive surface runoff, they construct a water-collecting & escape network in the form of open canals.

Waste ditches are arranged on natural depressions with maximum use of thalwegs by the beeline to the water intake along existing roads regardless of side ditches and by side ditches, along water-use boundaries, and along distribution canals.

On-farm waste water is diverted to the farm's tail ditches, and the water from the latter is diverted to the main tail ditch. The least distance between two waster ditches is taken as 800-1200 m and twice as much at double-sided command of distributors.

Irrigation ditches with a flow rate over 250 l/s end not in dead end but in escape structures through which water comes to waste ditches.

In large inter-farm distributors, main canals and their branches, except for tail escapes, emergency distributors are made. If large irrigation canals are arranged across natural slopes, interception drains are made on their upstream parts, which serve for catching flood and storm waters.

Design and calculation of ditches. Water-collecting & escape ditches are built in groves and, as a rule, with trapezoidal section. The elements of ditch cross-section are determined by hydraulic calculation. The bottom width is calculated and taken as 0.3 m at least, and depth as 0.8-1 m. At maximum flow rates, the flow velocity in ditches should be lower than the erosive velocity and higher than the silting velocity so that to avoid erosion, silting, and weeding of the ditches.

At design flow rates, the water level in waste ditches is to be 15-20 cm above the ground surface, making provision for discharge of surface runoff from the lowest sites. The water level in a higher-order water-collecting & escape ditch should be at least 5 cm lower the water level in a lower-order waste ditch.

The design flow rate of tail escapes for earth canals are taken as 0.25-0.5 rated flow rate of the permanent irrigation canal at the end site. The design flow rate of collecting ditches is taken of up to 30% of the gross rated flow rates of simultaneously operating irrigation canals that discharge water into this water-collection canal.

Discharge/outlet works to drain water from irrigation canals at the end of the distributors and in the places of disaster discharge. Bridges and pipelines are constructed in the places of crossing of the waste ditches with roads; pipelines and inverted siphons (pipe canals) at the places of crossing with irrigation canals; and check drops and inclined drop structures at the steep-gradient sites.

Lock-weir and Drainage Network

Lock-weir and drainage network is built up to protect against flooding and waterlogging and possible salinization of soils on irrigated areas. In combination with the irrigation network, it provides bilateral control of the soil water regime.

Lock-weir and drainage network includes:

- intercepting drainage system canals, i.e. interception drains/ditches, interception and catch-water drains, catch-water drains, which prevent entry of surface and ground waters to the irrigated site;

- main discharge/wasteway canal, or collector, laid at the lowest levels of the irrigated area and diverting discharge and drainage waters from the irrigated land;

- inter-farm escape/wasteway canal, or collector, which receives and diverts discharge and drainage waters from farms' territories;

- farm escape/wasteway canal, or collector, which receives and diverts discharge and drainage waters from the territory of a single farm;

- inter-plot or inter-brigade escape/wasteway canal, or collector, which receives and diverts discharge and drainage waters from the territory attached to crop-rotation or brigade land plots;

- plot or brigade escape/wasteway canal, or collector, which receives and diverts discharge and drainage waters from the territory attached to a single crop-rotation or brigade land plot;

- small-scale lock-weir network: irrigation plan, check, plot organized at small plots.

Man-made Structures

Man-made structures represent hydraulic facilities on all canals to regulate and control water in the system.

Man-made hydraulic structures on irrigation system canals are needed to control water flow rate and level in particular parts of the system as well as the velocity and quality of the water delivered. Among water conveyance/passageway structures are flumes, tunnels, aqueducts, inverted siphons (pipe canals), and inclined drop structures.

Flumes are man-made channels with various cross-sections; their bottom is placed on the ground or on special supports, trestle bridges. Water motion there is generally of free flow nature. Aqueducts represent bridge-type hydraulic structures made from precast or cast-in-situ reinforced concrete, which form man-made channels (dikes) above surface depressions (ravines, roads, inverted siphons/pipe canals, etc.). The most important place is the place of connection of the aqueduct with a canal. Here, erosion, seepage, and destruction of canal sections may easily happen. Therefore, pile retaining and sheet-pile walls cut into the bottom and side slopes to a depth of not more than 0.8-1.0 m.

Inverted siphons are hydraulic structures that represent water conduits serving to pass water under a channel or when crossing a deep valley or road. The pressure required for water to move is created due to the difference in the inlet and outlet holes. The Inverted siphon can be with one or several pipes.

Water supply is regulated by means of offtake regulators, which also regulate water elevation and velocity (being equipped with flow meters). There are offtake regulators of open (man-made channel/flume bounded by sidewalls and equipped with gates of different design) and closed (pipes) types.

Check drops are conjugation structures of open or closed design for passing water and dissipating energy at a short distance. There are cascade-type (open-conduit drop, baffled outlet drop, and closed conduit drop) and cantilever-type drops.

Inclined drop structures are man-made channels with gradient that is more than the critical which connect different-level canal sections.

Among control structures are spillways and water outlets as well. They are installed both at the head and tail sections of the canal. They are designed for water discharge in case of overfilling and for water evacuation. Spillways are installed on permanent canals.

Water outlets (fixed and portable) serve for water delivery from permanent canals to temporary irrigation ditches.

The following structures are placed in the closed network: distribution pots (for regulating water supply) with manual or automatic control; irrigation valves; sinks (in depressions); air valves (in highest points of the route); dashpots (in the pressure pipeline after a valve; at each 2-3 km along the pipeline).

Man-made Structures in the Irrigation Network

The following hydraulic facilities are installed in the irrigation network in order to provide water control, measurement, and management: water outlets; water-retaining structures; grade-control structures; and water conveyance structures (aqueducts).

The hydraulic facilities in the irrigation network are broken down into cast-in-place, precast, and combined types.

Cast-in-place structures are built from concrete, reinforced concrete, stone, wood, while precast ones from separate fabricated blocks.

Combined structures are installed on site, and their particular parts are mounted from prefabricated blocks.

Structures on canals can be of closed (piped) and open types.

Water outlets (regulators) are mounted in the head of permanent irrigation canals (distributors) and on temporary irrigation ditches. Water outlets are built with and without crossing. If the canal route has steep slope, water outlets are combined with check drops or inclined drop structures. To regulate water level and flow rate, water outlets are equipped with plain or sector gates. Pipe outlets (closed-conduit waterways) have usually round and sometimes square or rectangular cross-section.

Water-retaining structures (check dams) are arranged on inter-farm and large farm canals. The purpose of these structures is to keep required water level in canal at water discharge below Qnorm.

Grade-control structures are installed at the places with considerable slope of canal route. Flow velocity in the earth canal will exceed admissible scouring rates. Inclined drops, cantilever-type (lip) drops, and shaft-type (cylinder) drops are used as grade-control structures.

Water conveyance structures are built in the places where water needs to be transported over some obstacles or where construction of earth canals would be difficult from the technical or economically impractical. Among such obstacles are gullies, ravines, rivers, deep valleys, canals, roads, etc. Inverted siphons, aqueducts, flumes, pipe culverts, and tunnels are qualified as water conveyance structures.

The inverted siphon is a pressure pipeline constructed where a canal crosses a natural or man-made obstacle. It consists of culvert inlet and culvert outlet as well as pressure pipeline. Most often factory-made reinforced-concrete, more rarely asbestos-cement or steel pipes are used for inverted siphons. At higher capacity, inverted siphons are usually made from cast in-situ reinforced concrete. The number of inverted siphon pipe sockets or runs depends on the discharge of the water passed. At the inverted siphon's culvert inlet, service or repair gates as well as trash-rack structures are installed. The flow velocity in inverted siphons varies from 1 to 4 m/s; it should be higher that the silting and average flow velocity.

The aqueduct represents an elevated flume. It consists of culvert inlet and culvert outlet as well as a flume with supports. At present, aqueduct flumes are as a rule made from cast in-situ reinforced concrete or precast reinforced concrete. Piles, rack-mountable frames with foundations, land abutments, etc. are used as supports. Flow velocity in aqueducts is set within the range of 1.0-2.5 m/s. Prefabricated circular culvert aqueducts made from reinforced concrete are installed at the places where an irrigation canal crosses bodies of roads or canals.

Open Channel Flow

There are two kinds of flow: Open channel and pipe. These flows differ in many ways. The important difference is that open channel flow has free water surface whereas pipe flow does not have free water surface.

Open Channels

Irrigation water is conveyed in either open channel or closed conduits. Open channels receive water from natural streams or underground water and convey water to the farm for irrigation. Open channels have free surface. The free surface is subjected to atmospheric pressure. The basic equations used for water flow in open channels are continuity equation, Bernoulli equation and Darcy Weisbach equation.

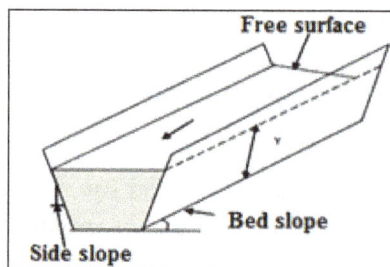

A trapezoidal shaped open channel.

Types of Open Channels

Prismatic and Non-Prismatic Channels

A channel in which the cross sectional shape, size and the bottom slope are constant over long stretches is termed as prismatic channel. Most of the man-made or artificial channels are prismatic channels. The rectangular, trapezoidal, triangular and half-circular are commonly used shapes in manmade channels. All natural channels generally have varying cross section and consequently are nonprismatic.

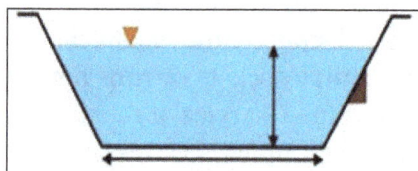

Sketch of a prismatic channel.

Rigid and Mobile Boundary Channels

Rigid channels are those in which the boundary or cross section is not deformable. The shape and roughness magnitudes are not functions of flow parameters. The lined channels and non erodible unlined channels are rigid channels. In rigid channels the flow velocity and shear stress distribution are such that no major scouring, erosion or deposition takes place in the channel and the channel geometry and roughness are essentially constant with respect to time. Channels are classified as mobile channels when the boundary of the channel is mobile and flow carries considerable amounts of sediment through suspension and is in contact with the bed. In the mobile channel, depth of flow bed width, longitudinal slope of channel may undergo changes with space and time depending on type of flow. The resistance to flow, quantity of sediment transported and channel geometry all depend on interaction of flow with channel boundaries.

Types of Open Channel Flow

Open channel flow can be classified into many types and described in various ways. The following section describes classification based on variation of flow properties such as depth of flow, velocity etc. with respect to time and space.

a) Steady and Unsteady Flows

Flow is steady if the velocity and depth are constant with respect to time. If the depth velocity or discharge changes with time, the flow is termed as unsteady.

Flood flows in rivers and rapidly varying surges in canals are examples of unsteady flow.

b) Uniform and Non-Uniform Flows

If the flow properties, say the depth of flow and discharge in an open channel remain constant along the length of the channel, the flow is said to be uniform. A prismatic channel carrying a certain discharge with a constant velocity is an example of uniform flow.

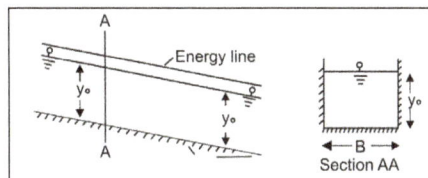

Uniform flow in a prismatic channel.

If the flow properties such as depth and discharge vary with distance along the channel is termed as non-uniform flow.

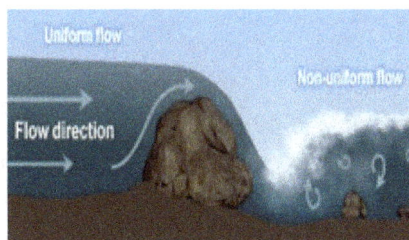

Uniform and non-uniform flows.

Figure shows a view of uniform and non uniform flow. In uniform flow, the gravity force on the flowing water balances the frictional force between the flowing water and inside surface of the channel, which is in contact with the water. In case of non-uniform flow, the friction and gravity forces are not in balance.

The flow in open channel can be steady or unsteady. It can be uniform or non -uniform. A non-uniform flow is also termed as varied flow. Steady and unsteady flows can be uniform or varied.

c) Gradually Varied and Rapidly Varied Flow

The non-uniform flow can be classified as gradually varied flow (GVF) and rapidly varied flow (RVF). Varied flow assumes that no flow is externally added to or taken out of channel system and hence the volume of water in a known time interval is conserved in the channel system and hence the volume of water in a known time interval is conserved in the channel system. If the change of depth is gradual so that the curvature of streamlines is not excessive, such a flow is said to be gradually varied flow (GVF).

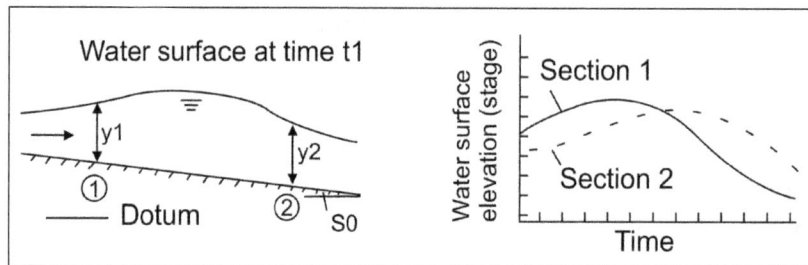

Gradually flow.

Figure above shows water surface profile of a GVF; here y1 and y2 are the depth at section 1 and 2, respectively. In GVF, the loss of energy is essentially due to boundary friction. Therefore, the distribution of pressure in the vertical direction may be taken as hydrostatic. If the curvature in a varied flow is large and the depth changes appreciably over short lengths, then the flow is termed as a varied flow. It is a local phenomenon. The examples of RVF are hydraulic jump and hydraulic drop.

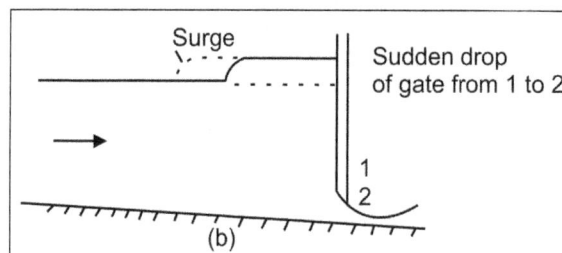

Rapidly varied flow.

d) Spatially Varied Flow (SVF)

Addition or diminution of water along the course of flow causes non uniform discharge and the resulting flow is known as spatially varied flow (SVF). Hydraulic behaviour of spatially varied flow with increasing discharge (case of drainage channel) is different in certain respects from that of spatially varied flow with decreasing discharge (in case of irrigation channel). Figures above shows a case of spatially varied flow with decreasing discharge. Figure shows case of increasing discharge.

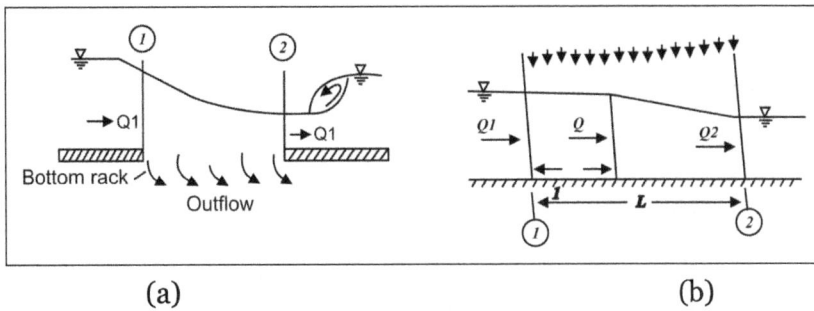

(a)   (b)

i) Spatially Varied Flow with Increasing Discharge

In this type of spatially varied flow, an appreciable portion of the energy loss is due to the turbulent mixing of the added water and the water flowing in the channel. In the most cases, this mixing is of relatively high magnitude and uncertainty. Because of the resulting high and uncertain losses, the momentum equation is more convenient than the energy equation in solving spatially varied flow with increasing discharge. From a practical viewpoint, the high energy loss seems to make channels designed for such spatially varied flow hydraulically inefficient, but physical circumstance and economical considerations sometimes make the use of such channels desirable.

ii) Spatially Varied Flow with Decreasing Discharge

Fundamentally, these types of spatially varied flow may be treated as a flow diversion where the diverted water does not affect the energy head. This concept has been verified by both theory and experiments. Therefore, the use of the energy equation is more convenient in solving spatially varied flow with decreasing discharge. The theory of spatially varied flow with decreasing discharge was probably employed first in the design of lateral spillways or side spillway weirs. This type of structure is usually a long notch installed along the side of a channel for the purpose of diverting or spilling excess flow.

The spatially varied flow with decreasing discharge is encountered in the design of irrigation water conveyance system whereas with increasing discharge in design of surface and subsurface drainage systems.

State of Flow

The state of flow in open channels is influenced by viscosity, gravity and inertial forces. The ratio of inertial to viscous force is the Reynolds number.

Effect of Viscosity

In open channel flow may be laminar, transitional or turbulent depending on viscosity in relation to inertial force. If viscous forces are strong in comparison to inertial force, the flow can be laminar otherwise vice versa for turbulentant flow.

The characteristic length-scale for an open channel of width (b) and depth (y), the hydraulic radius (R) = (by/b+2y). As a general rule, open channel flow in laminar, if Reynolds number defined by

$$\text{Re} = \frac{VR}{v} \text{ is less than 500.}$$

Where,

>   V = Flow Velocity

>   R = Hydraulic radius

>   v = Kinematic viscosity.

In open channels the transitional range of Re for practical purpose, is considered between 500 and 2000. The revalue exceeding 2000 is considered as turbulent flow.

In close conduits the flow is i) laminar for Re < 2000, ii) transitional 2000 < Re < 4000 and iii) turbulent Re > 4000.

Effect of Gravity

The effect of gravity is represented by ratio of inertial forces to gravitational forces. This ratio is known as Froude number (Fr), given by

$$Fr = \frac{V}{\sqrt{gL}}$$

>   V = mean flow velocity,

>   g = acceleration due to gravity,

>   L = characteristic length (it can be hydraulic depth, y or hydraulic radius, R).

For the flow to be critical (Fr = 1) i.e.

$$V = \sqrt{gy}$$

For sub critical flow (Fr < 1) i.e

$$V \langle \sqrt{gy} \text{ and}$$

For super critical flow (Fr > 1) i.e.

$$V \rangle \sqrt{gy}.$$

Seepage in Canals and Field Channels

Seepage loss in unlined canals and farm ditches often range from one-fourth to one-third of the total water diverted. In extremely sandy or gravelly ditches, half the water can be lost through seepage. Reducing seepage by using improved conveyance facilities can increase water available for crop needs, allowing irrigation of additional land, prevent water-logging, increase in channel capacity, reduction in maintenance cost and more importantly enable to use available water sustainably. Especially in the regions of water scarcity, minimising the seepage losses is important.

Measurement of Seepage in Canal

The most commonly used methods applied for measuring the quantity of water lost due to seepage in a canal section are as follows:

i)   Ponding method,

ii) Inflow-outflow method,

iii) Seepage meter method.

Ponding Method: The ponding method is one of the simplest methods of determining seepage from a canal section. Water is ponded in temporary water-tight dikes constructed in a straight length of canal under investigation. The time rate of drop of the water in the canal level is measured. The dimension of the ponded reach of the canal are measured the seepage computed as volume of water lost from the canal per unit wetted area of canal per unit time and normally is expressed as $m^3/m^2/day$.

Inflow-Outflow Method: The inflow-outflow method is based on measuring the rates of water flowing into and out of selected section of canal reach. It is based on water balance approach considering the inflows outflows and losses into account. Canal sections with minimum number of outlets and diversions and no appreciable inflow from higher lands are considered for seepage measurement. Water stage recorders are also used to record the height of flow in the flume as a function of the elapsed time. The seepage is computed as the difference in inflow and outflow per unit wetted area of canal section under consideration.

Seepage Meter Method: The seepage meter is a device for directly measuring the flow between ground water and surface water body such as lake or stream. The seepage meter is a modified form of the constant head permeameter. It is mainly used to determine location of relatively high seepage losses. Seepage meter can be constructed from inexpensive material such as galvanised iron sheet. Seepage meters are suitable when many measurements are needed to characterize groundwater surface water exchange in different sequent of water body.

Materials for Lining Canals and Field Channels

A large variety of lining materials for seepage loss control from canals and field channels is available for use. Lining of canals or channels offers other advantage such as enhance stability, increasing life, protection from flood in addition to seepage control. The various types of channel lining material commonly used are as follows:

i)   Hard surface linings:

   a)   Cement concrete or pre cost concrete,

   b)   Stone masonry,

   c)   Brick tile or concrete tile,

   d)   Asphaltic concrete.

ii) Earth type lining:

    a) Compacted earth,

    b) Soil cement,

    c) Bentonite - clay soil mixture.

iii) Synthetic sheet/film:

    a) Rubber or synthetic materials,

    b) Low density polyethylene sheet.

The following points are normally considered for selecting method of lining and materials.

    i) Availability and cost of the material at the site or within reach.

    ii) Labour available for lining at a reasonable cost.

    iii) Degree of water-tightness desired.

    iv) Velocity of flow in the channel.

    v) Useful life of the lining material.

    vi) Maintenance cost.

Design of Open Channel

Open channel design involves determining cross-section dimensions of the channel for the amount of water the channel must carry (i.e., capacity) at a given flow velocity, slope and, shape or alternatively determining the discharge capacity for the given cross-section dimensions.

The terminologies used in the design of open channels of different geometry are given below:

i) Area of Cross Section (a): Area of cross section of for a rectangular cross section, of wetted section. For a rectangular cross section, if b = width of channel and y = depth of water, the area of wetted section of channel (a) = b.y.

ii) Wetted Perimeter (p): It is the sum of the lengths of that part of the channel sides and bottom which are in contact with water. The wetted perimeter (p) = b+2y.

iii) Hydraulic Radius (R): It is the ratio of area of wetted cross section to wetted perimeter. The hydraulic radius,

$$(R) = \frac{a}{p} = \frac{by}{b+2y}$$

iv) Hydraulic Slope (S): It is the ratio of vertical drop in longitudinal channel section (h) to the channel length (l). Hydraulic slope

$$(S) = \frac{h}{I}$$

v) Freeboard: It is the vertical distance between the highest water level anticipated in channel flow and the top of the retaining banks. This is provided to prevent over topping of channel embankments or damage due to trampling. This is provided between 15.25% of normal depth of flow.

Discharge Capacity of Channel

Channel capacity can be estimated by equation given as:

$$Q = \frac{(16667)(\text{DDIR})(A)}{(\text{HPD})(E_i)}$$

where,

DDIR = design daily irrigation requirement (mm/day)

A = irrigated area supplied by canal or ditch (ha)

HPD = hours per day that water is delivered

E_i = irrigation efficiency including conveyance efficiency of canal or ditch (percent).

The velocity of flow in a canal or ditch should be non erosive and non silting that prevent the deposition of suspended substances. Normally flow velocity in excess of 0.6 m/s is non silting. The maximum velocity that does not cause excessive erosion depends on the erodibility of the soil or lining material. The maximum allowable velocities for lined canals and unlined ditches listed in table can be used when local information is not available.

Economical Section of a Channel

A channel section is said to be economical when the cost of construction of the channel is minimum. The cost of construction of a channel depends on depth of excavation and construction for lining. The cost of construction of channel is minimum when it passes maximum discharge for its given cross sectional area. It is evident from the continuity equation and uniform flow formulae that for a given value of slope and surface roughness, the velocity of flow is maximum when hydraulic radius is maximum. The hydraulic radius is maximum for given area if wetted perimeter is minimum. Hence the wetted perimeter, for a given discharge should be minimum to keep the cost down or minimum. This condition is utilized for determining the dimensions of economical sections of different forms of channels. Most economical section is also called the best section or hydraulic efficient section as the discharge passing through a most economical section of channel for a given cross-sectional area (A), slope of the bed (So) and a roughness coefficient (n), is maximum.

The conditions for the most economical section of channel:

i)   A rectangular channel section is the most economical when either the depth of flow is equal to half the bottom width or hydraulic radius is equal to half the depth of flow.

ii) A trapezoidal section is the most economical if half the top width is equal to one of the sloping sides of the channel or the hydraulic radius is equal to half the depth of flow.

iii) A triangular channel section is the most economical when each of its sloping side makes an angle of 450 with vertical or is half square described on a diagonal and having equal sloping sides.

The discharge from a channel is given by,

$$Q = AV = AC\sqrt{RS_0} = AC\sqrt{\frac{A}{P}}\,S_0 = K*\frac{1}{\sqrt{P}}$$

where Q = discharge (m3/s), A = area of cross section (m2), C = Chezys constant,

R= Hydraulic radius (m), P = wetted perimeter (m), = bed slope (fraction or m/m), K = constant for given cross sectional area and bed slope and = A3/2 C So1/2.

In equation $Q = AV = AC\sqrt{RS_0} = AC\sqrt{\frac{A}{P}}\,S_0 = K*\frac{1}{\sqrt{P}}$ the discharge Q will be maximum when the wetted perimeter P is minimum.

i) Channel Shape: Among the various shapes of open channel the semi-circle shape is the best hydraulic efficient cross sectional shape. However the construction of semicircle cross section is difficult for earthen unlined channel. Trapezoidal section is commonly used cross section.

ii) Channel Dimensions: The channel dimensions can be obtained using uniform flow formula, which is given by,

Q = A V

Where,

V = flow velocity (m/s)

A = cross-sectional area of canal perpendicular to flow (m²)

Q = capacity of the channel (m³/s)

Velocity is computed by Manning's formula or Chezy formula.

Manning's Equation is given by,

$$V = \frac{1}{n}R^{2/3}S^{1/2}$$

Chezy's equation is given by,

$$V = CR^{1/2}S^{1/2}$$

Where,

n = Manning's roughness coefficient

C = Chezy's roughness coefficient

R = hydraulic radius (m)

S = bed slope (m/m).

Table. Limiting velocities for clear and turbid water from straight channels after aging.

Velocity and Water transporting		
Material	m/s	m/s
Fine sand, colloidal	0.46	0.76
Sandy loam, noncolloidal	0.53	0.76
Silt loam, noncolloidal	0.61	0.92
Alluvial silts, noncolloidal	0.61	1.07
Ordinary firm loam	0.76	1.07
Volcanic ash	0.76	1.07
Stiff clay, very colloidal	1.14	1.52
Alluval silts, colloidal	1.14	1.52
Shales and hardpans	1.83	1.83
Fine gravel	0.76	1.52
Graded loam to cobbles when noncollodal	1.14	1.52
Graded silts to cobbles when colloidal	1.22	1.68
Coarse gravel, noncolloidal	1.22	1.83
Cobbles and shingles	1.53	1.68

Energy Depth Relationship

From hydraulic point of view, the total energy of water in any streamline passing through a channel section may be expressed as total head, which is equal to sum of the elevation above a datum, the pressure head, and the velocity head. The total energy at the channel section is given by,

$$H = z + y + \frac{v^2}{2g}$$

where,

H = total energy,

z = elevation head above datum,

y = depth of water in channel,

V = velocity of flow,

g = acceleration due to gravity.

The specific energy is the total energy at any cross section with respect to channel bed. Considering slope of the channel bed is very small, the specific energy E is,

$$E = y + \frac{v^2}{2g}$$

For the channel of rectangular section having width b, the cross sectional area of channel.

$$A = b \, y$$

then

$$E = y + \frac{Q^2}{2gb^2y^2}$$

Differentiating equation $E = y + \dfrac{Q^2}{2gb^2y^2}$, equating it to zero for minimum condition, this becomes:

$$\frac{dE}{dy} = 1 - \frac{Q^2dA}{gA^3dy} = 1 - \frac{v^2}{gA}\frac{dA}{dy};$$

$$\text{but } \frac{dA}{dy} = b$$

$$\text{Hence } \frac{dE}{dy} = 1 - \frac{Vc^2}{gy_c} = 0$$

When V Vc, Y = (Critical depth),

$$\frac{V^2c}{gy_c} = 1$$

$\dfrac{V}{\sqrt{gy_c}}$ is defined as Froude number, for flow to be critical its value is equal to 1. It is greater than 1 for super critical flow and less than 1 for sub critical flow.

Critical depth (Y_c) for rectangular channel is given by,

$$y_c = \left[\frac{Q^2}{gb^2} \right]^{1/3}.$$

The principle of design of flumes and hydraulic structures (open drop and chute spillways) is based on the concept of specific energy and critical flow.

Velocity Distribution in a Channel Section

The velocity of flow in any channel section is not uniformly distributed. The non- uniform distribution of velocity is due to the presence of a free surface and the frictional resistance along the channel surface. In a straight reach of channel section, maximum velocity usually occurs below the free surface at a depth of 0.05 to 0.15 of the total depth of flow. The velocity distribution in a channel section depends on various factors such as the shape of the section, the roughness of the channel and the presence of bends in the channel alignment. The man velocity of flow in a channel section can be computed from the vertical velocity distribution curve obtained by actual measurements. It is observed that the velocity at 0.6 depth from the free water surface or average of the velocities measured at 0.2 depth and 0.8 depth from free water surface which is very close to the mean velocity of flow in the vertical section. The velocity can be measured by pitot tube or current meter.

Irrigation Scheduling for Regulated Deficit Irrigation

Irrigation is generally associated with minimising moisture stress. Under such conditions trees grow quickly and are very vigorous. Until a tree has reached its desired size it should not be stressed for water. Once the tree has grown to its desired size, however, vigorous growth not only increases the need for pruning but can reduce yield. Irrigation needs to be managed in such a way as to control the growth of shoots. Such management is known as regulated deficit irrigation (RDI) and in experimental plots has maintained yields of pears and peaches, and reduced irrigation by about 30%.

The RDI Technique

With RDI, trees are kept short of water when fruit growth is slow or after harvest but are given ample water during the time of rapid growth of fruit. This reduces the growth of shoots. If RDI is properly managed, there is no reduction in the size of fruit or yield. The reason why the above technique works relates to the growth pattern of shoots and fruit. On most deciduous fruit trees, the shoots grow rapidly early in the season and their growth slows down as the fruit begins to grow rapidly. In contrast, early in the season the fruit grows slowly. Water stress at this time will reduce the growth of shoots without markedly affecting the growth of fruit.

With RDI, the irrigation season can be divided into four periods. The duration of these periods is determined by both weather and the relationship between vegetative growth and the growth of fruit.

Period 1

During this period immediately following flowering, care needs to be exercised to avoid water stress particularly in stone fruit. For example, in peaches there is an initial rapid fruit growth for approximately 4 weeks following flowering when the soil should not be allowed to dry out beyond 40 kPa in sandy soil and 60 kPa in clay loam soils.

In most seasons in the Goulburn Valley, crops like pears are not irrigated until reference crop evapotranspiration (ET_o) exceeds rainfall by 125 mm. Generally, this is in late October but could be as late as mid-November in a wet spring. However, in recent years there has been insufficient winter and early spring rain to wet up the root zone. Root zone soil moisture must be measured to avoid water stress. Similarly, in environments dissimilar from the Goulburn Valley (for example, trees growing in lighter soil types) measurements of soil moisture will avoid the root zone drying out excessively.

Period 2

Period 2 commences approximately four to five weeks after flowering and continues until six weeks before harvest for early-maturing fruits (that is, before mid-January), and eight weeks before harvest for later maturing fruits. Trees are irrigated with greatly reduced volumes of water compared to that which would normally be applied. Irrigation to replace 30 % of orchard water use capability is recommended. Soil moisture in the middle of the wetted fibrous root zone should not exceed 100 kPa in sand or 400 kPa in clay loams.

Period 3

In this period six to eight weeks before harvest, the fruit is growing rapidly and the tree now needs ample water to reach maximum fruit size. Water stress must not occur during this final period of fruit growth. Irrigation to replace 100% of orchard water use capability is recommended. Soil moisture in the middle of the wetted fibrous root zone should not exceed 40 kPa in sand or 60 kPa in clay loams.

Period 4

After harvest a similar strategy as during period 2 can be implemented. In early maturing varieties and species (for example, cherries and apricots) there is considerable shoot growth after harvest which should be kept in check to maintain fruitfulness and even cropping within the canopy. Irrigation to replace 30 % of orchard water use capability is recommended. Soil moisture in the middle of the wetted fibrous root zone should not exceed 100 kPa in sand or 400 kPa in clay loams.

Scheduling RDI from ET_o

In all three periods, reference crop evapotranspiration (ET_o) readings, which are readily available in most districts, can be used to schedule irrigation. However, it is strongly recommended that soil moisture monitoring be integrated into an irrigation schedule to avoid over- or under-irrigating trees.

In table, examples of how to use ET_o to schedule RDI in a peach orchard are shown for drip, microjet and sprinkler irrigation. The table is divided vertically into three sections; each section refers to a different form of irrigation - drip, microjet and sprinkler.

To show the influence that the spacing between trees has on the calculations for scheduling of irrigation, different spacings between trees are used for each of the three systems of irrigation. As previously mentioned, the irrigation season is divided into three periods, and the calculations needed during each of these periods are set out below the appropriate period. These calculations are divided into various sub-headings shown on the left side of table.

Weekly ET_o

Values for weekly ET_o (mm) can be obtained from the Bureau of Meteorology. Data shown in the table are typical for the Goulburn Valley. To use the table you merely have to replace the figures given in the example by those that you have collected in the previous week.

Effective Area of Shade

Orchard effective area of shade (EAS, %) is a simple and practical estimate of tree size and hence the actively transpiring leaf area in an orchard. EAS is determined from measurements of the percent shade cast by the trees at three key times a day ($3\frac{1}{2}$ h before solar noon, at solar noon and $3\frac{1}{2}$ h after solar noon). Taking three measurements per day accounts for differences in foliage extent (i.e. training system and tree size), planting arrangement (i.e. row orientation and tree spacing) and leaf area density. EAS is calculated from the average of the three measurements. The percent shade can be estimated visually or measured using a light bar known as a ceptometer.

Understorey Coefficient

The understorey coefficient (K_e) is a factor to convert ET_o to understory water use; the combination of soil evaporation and cover crop water use. For modern high-density orchards, micro-irrigation is designed to deliver water requirements to individual trees and minimize the contribution of irrigation to understorey water use. Hence under drip irrigation K_e can be set to 0.1 and under microjet set to 0.2. Whereas under sprinkler irrigation there is substantial understorey water use and K_e can be set to 1-EAS. For example, if EAS = 20 % then K_e = 1-20 % = 0.8.

Stress Coefficient

The stress coefficient (K_s) is a factor used for setting the amount of stress deliberately imposed on the orchard. A value of 1.0 is no stress. For example, during period 2 under RDI it is recommended to replace 30 % of orchard water use capability, hence K_s = 0.3.

Weekly Orchard Irrigation

Weekly irrigation for a peach orchard (I) is calculated from weekly ET_o, effective area of shade (EAS), the understorey coefficient (K_e) and the stress coefficient (K_s) using the following formula:

$$l = k_s \times ET_0 \times \left[(1.1 \times EAS) + k_s \right]$$

Area of Planting Square

The area of planting square (m ^{2}) is calculated from the distance between rows multiplied by the in-row distance between trees. Different spacings between trees are given for each form of irrigation in the example in Table.

Weekly Tree Irrigation

Weekly irrigation requirement per tree is calculated from the weekly orchard irrigation multiplied by the area of planting square.

Recommended Interval between Irrigations

The interval between irrigations (day) is also important with RDI, and recommended intervals are given in Table. For drip irrigation, the rationale behind these recommendations relates to the size of the wetted root zone. In period 2, frequent irrigation (that is, daily) wets a small volume of soil regularly. In contrast, using a two-day interval in period 1 and 3 enables a much greater volume of water to wet a much larger root zone. This manipulation in wetting the root zone could be responsible for the observed improved growth of fruit in period 3 and higher yields on RDI-managed trees. If, with drip irrigation, the system has to be run for more than 24 hours every second day to provide the required quantity of water, serious thought should be given to upgrading the system to a higher rate of discharge.

The longer interval between irrigations in period 2, than in period 1 and 3, for both microjet and sprinkler irrigation is necessary to allow sufficient water to wet the soil to a reasonable depth.

In period 2, with microjet and sprinkler irrigation, an interval of seven and 21 days respectively is recommended. If the combined effects of evaporation, spacing of trees and rate of application result in less than two- and eight-hour irrigation times respectively for microjet and sprinkler irrigations, the interval will need to be extended until such figures are reached. For these long intervals, irrigation is based on the accumulated evaporation since the previous irrigation.

Table: Example calculations of irrigation interval and run time for RDI under drip, microjet and sprinkler irrigation.

	Drip 6 m x 1 m planting			Microjet 5 m x 3 m planting			Sprinkler 6 m x 6 m planting		
	Period 1	Period 2	Period 3	Period 1	Period 2	Period 3	Period 1	Period 2	Period 3
Weekly ET o (mm)	20	35	45	20	35	45	20	35	45
Effective area of shade (%)	30	60	60	30	60	60	20	50	50
Understorey coefficient	0.1	0.1	0.1	0.2	0.2	0.2	0.8	0.5	0.5
Stress coefficient	1	0.3	1	1	0.3	1	1	0.3	1
Weekly orchard irrigation (mm)	8.6	8.0	34.2	10.6	9.0	38.7	20.4	11.0	47.3
Area of planting square (m^2)	6	6	6	15	15	15	36	36	36
Weekly tree irrigation (litre/tree)	52	48	205	159	136	581	734	397	1701
Recommended interval between irrigation (day)	2	1	2	4	7	3	5	21	5
Water required at each irrigation (litre/tree)	15	7	59	91	136	249	525	1191	1215
Application rate (litre/h/tree)	8	8	8	40	40	40	120	120	120
Run time (hour)	2	1	7½	2¼	3½	6¼	4¼	10	10¼

Water Required at Each Irrigation

The quantity of water required at each irrigation is multiplied by the interval between irrigations in days and divided by 7 (that is, by the number of days in the week). For example, if the weekly irrigation requirement is 52 litre but the interval is only two days, then approximately 15 litre of water is applied every 2nd day (52 x 2÷7 = approximately 15).

Application Rate

Application rate (litre/hour/tree) is the amount of irrigation applied to each tree per hour. This is calculated from the emitter discharge rate multiplied by the number of emitters per tree. If not known, this should be measured.

Run Time

Run time (hour) is calculated by dividing the number of litres per tree required at each irrigation by the application rate.

RDI with Flood and Furrow Irrigations

With surface irrigations, such as flood or furrow, it is difficult to control the amount of water applied per irrigation. Nevertheless, the principles discussed above apply; the initial irrigation can be delayed and the interval between irrigations can be increased in period 2. After 12 years of experimenting with RDI it became obvious that in the past, much water was wasted on early irrigation.

Irrigation System Controllers

A controller is an integral part of an irrigation system. It is an essential tool to apply water in the necessary quantity and at the right time to sustain agricultural production and to achieve high levels of efficiency in water, energy and chemical uses.

Irrigation controllers have been available for many years in the form of mechanical and electromechanical irrigation timers. These devices have evolved into complex computer-based systems that allow accurate control of water, energy and chemicals while responding to environmental changes and development stages of the crop.

Basic Control Strategies

Two general types of controllers are used to control irrigation systems: Open control loop systems, and closed control loop systems. The difference between these is that closed control loops have feedback from sensors, make decisions and apply decisions to the irrigation system. On the other hand, open control loop systems apply a preset action, as is done with irrigation timers.

Open Control Loop Systems

When using an open control loop system, a decision is made by the operator or the amount of water and the time at which this water should be applied. The operator then goes on to set an irrigation controller according the desired schedule. These devices require external intervention they are referred to in control terms as open loop systems.

Open loop control systems use irrigation duration or applied volume for control purposes. Figure 1 shows the basic components of an open loop time- based irrigation controller. Notice that in this type of controller the basic control parameters are how often and how long is irrigation water is to be applied. Open loop controllers are also constructed in such a way that a clock is used to start irrigation and the application of a given volume to stop irrigation. In this type of controller the parameters set by the system operator are how often and the volume of water to be applied.

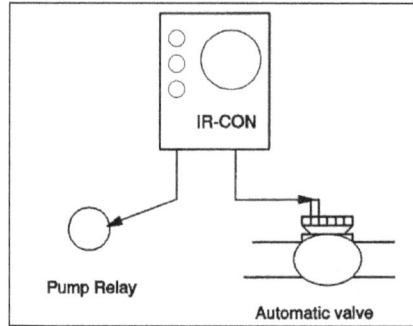

Open loop control systems have the advantages that they are low cost, readily available, and many variations of the devices are manufactured with different degrees of flexibility related to the number of stations and schedule specification. However, they do not respond automatically to changing conditions in the environment and require frequent resetting to achieve high levels of irrigation efficiency.

Closed Control Loop Systems

In a closed control loop the operator sets up a general strategy for control. Once the general strategy is defined, the control system takes over and makes detailed decisions of when to apply water and how much water to apply. This type of system requires that feedback be given back to the controller by one or more sensors. Depending on the feedback of the sensors, the irrigation decisions are made and actions are carried out if necessary. It is important to note that in this type of systems the feedback and control of the system is done continuously. Figure shows the elementary components of this type of system.

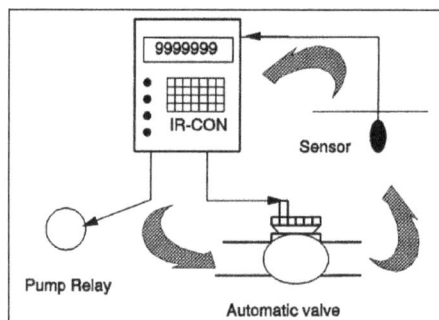

Closed loop controllers require data acquisition of environmental parameters, such as, soil-moisture, temperature, radiation, wind-speed and relative humidity. The state of the system (for example measured soil-moisture using a sensor as illustrated in figure) is compared against a desired state and a decision based on this comparison is made whether irrigation should be applied or not. Closed loop controllers for irrigation systems base their irrigation decisions on: 1) direct measurement of soil-moisture using sensors, 2) calculations of water used by the plants based on climatic parameters, or 3) both soil moisture sensors and climatic parameter measurements.

When using a computer-based controller, a very important component of a closed loop control system is the logic that is used to make decisions about operation of the irrigation system. Some of these systems may be very elaborate and use complicated simulation models that are verified with soil moisture measurements to arrive at an irrigation decision and implement the action at the appropriate time. Systems of this type (with different levels of complexity), are quickly being developed, and some have become commercially available in the past few years.

The simplest form of a closed loop control system is that of a high frequency irrigation controller that is interrupted by a moisture sensor. The sensor in figure is wired into the line that supplies power from the controller to the electric solenoid valve. The sensor operates as a switch that responds to soil moisture. When sufficient soil-moisture is available in the soil, the sensor maintains the circuit open. When soil-moisture drops below a certain threshold, the sensing device closes the circuit, allowing the controller to power the electrical valve.

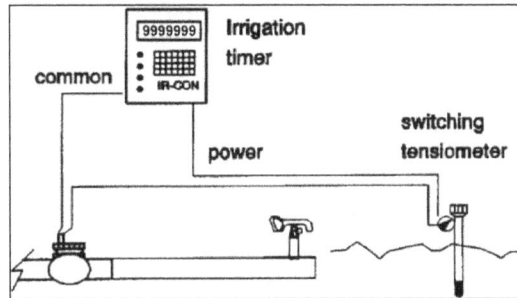

Using their arrangement in figure, the controller can be set to irrigate at a very high frequency (4 or 5 times more often than required). When the controller attempts to irrigate, irrigation will occur only if the soil-moisture sensor allows it, which in turn occurs only when soil-moisture has dropped below acceptable levels.

The system has been used successfully in controlling small sprinkler irrigated turf and microirrigated citrus at a research site using switching tensiometers. For turf, the tensiometers were installed at the center of the bottom third of the root system (10 inches deep) and the threshold was set to the point at which water stress symptoms were visible. In citrus a bank of sensors was used under the emitter connected in parallel, in such a way that any of the sensors would allow irrigation to occur.

The feedback system in figure is very low cost and is easy to install and maintain. However, the system has limitations:

1. Determining the best location of the sensor is not a straightforward task and requires some knowledge of soil-water and root dynamics,

2. Spatial variability of soil properties may result in readings that are not representative of the system.

Irrigation Timers

Irrigation timers are simple controllers consisting of clock units capable of activating one or more subunits of the irrigation system at specified times. Several designs are commercially available with many different features and over a wide range of costs.

By today's standards most irrigation timers provide several of the following functions:

- A clock/timer: Provides the basic time measurements by which schedules are executed.

- A calendar selector: This function allows definition of which days the system is to operate.

- Station time setting: This function allows definition of start time and duration for each station.

- Manual start: This function allows the operator to start the automatic cycle without disturbing the preset starting time.

- Manual operation of each station: This allows the operator to manually start the irrigation cycle without making changes to the preset starting time.

- Master switch: This function prevents activation of any station connected to the timer.

- Station omission: This function is used to omit any specified number of stations from the next irrigation cycle.

- Master valve control: Provides control to a master system valve. This function is used with certain types of backflow prevention equipment and also prevents flow to the system in case of failure in the system.

- Pump start lead: This feature allows a pump start solenoid to be activated whenever a station is activated, thus tying pump control with irrigation control.

The two most common types of controller designs are electromechanical and electronic.

Electromechanical Controllers

Electromechanical controllers use an electrically driven clock and mechanical switching (gear arrays) to activate the irrigation stations. These types of controllers are generally very reliable and not too sensitive to the quality of the power available. They generally are not affected by spikes in the power, and unless surges and brownouts are of such magnitude that they will damage the motor, they will continue to operate. Even if there is a power outage, the programmed schedule will not be lost and is generally delayed only for the duration of the power outage. However, because of the mechanically-based components they are limited in the features they provide.

Electronic Controllers

Electronic controllers rely on solid state and Integrated circuits to provide the clock/timer, memory and control functions. These types of systems are more sensitive to powerline quality than electromechanical controllers and may be affected by spikes, surges and brownouts. Particularly spikes and surges are common in rural areas in Florida where lightning tends to be frequent and intense. These types of systems may require electrical suppression devices in order to operate reliably. Because of the inherent flexibility of electronic devices, these controllers tend to be very flexible and provide a large number of features at a relatively low cost.

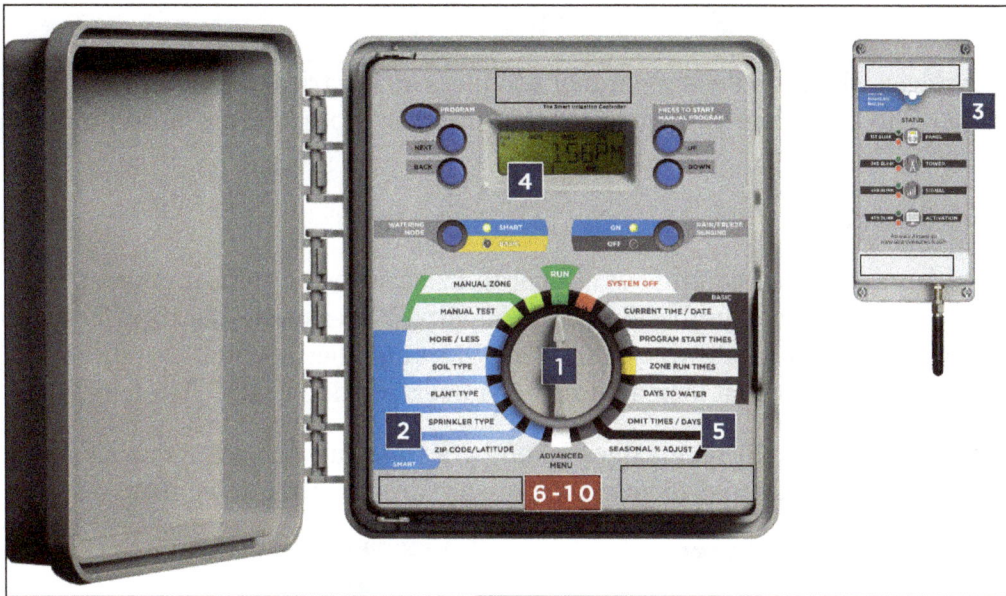

Computer-based Irrigation Control Systems

A computer-based control system consists of a combination of hardware and software that acts as a supervisor with the purpose of managing irrigation and other related practices such as fertigation and maintenance. This is done by the use of a closed control loop. A closed control loop consists of:

1. Monitoring the state variables,

2. Comparing the state variables with their desired or target state,

3. Deciding what actions are necessary to change the state of the system, and

4. Carrying out the necessary actions. Performing these functions requires a combination of hardware and software that must be implemented for each specific application.

Hardware Components

Figure shows the basic components of a closed loop control system, each of the hardware elements is described below.

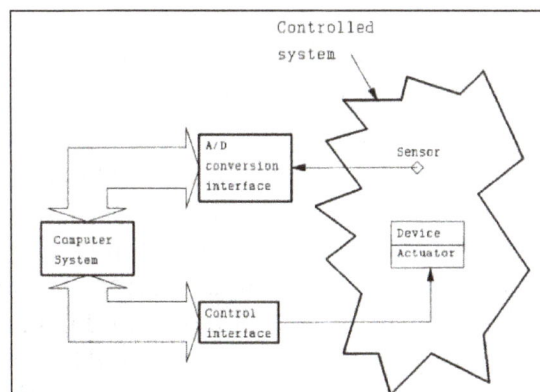

Sensors

A sensor is a device placed in the system that produces an electrical signal directly related to the parameter that is to be measured. In general, there are two types of sensors, continuous and discrete:

a) Continuous Sensors

Continuous sensors produce a continuous electrical signal, such as a voltage, current, conductivity, capacitance, or any other measurable electrical property. For example, sensors of different kinds can be used to measure temperature, such as thermistors and thermocouples. A thermocouple will produce a voltage difference that increases as the temperature increases.

Continuous sensors are used where values taken by a state variable are required and an on/off state is not sufficient, for example, to measure pressure drop across a sand filter.

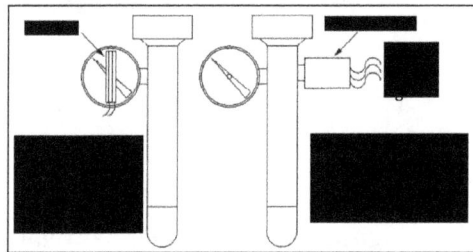

A tensiometer is used to measure water potential in soils. This illustrates
how it can be used as a discrete or continuous sensor.

b) Discrete Sensors

Discrete sensors are basically switches, mechanical or electronic, that indicate whether an on or off condition exists. Discrete sensors are useful for indicating thresholds, such as the opening and closure of devices (vents, doors, alarms, valves, etc.). They can also be used to determine if a threshold of an important state variable has been reached. Some examples of discrete sensors are a float switch to detect if the level in a storage tank is below a minimum desirable level, a switching tensiometer to detect if soil moisture is above a desired threshold, and a thermostat to indicate if a certain temperature has been reached. When combined with time, pulses from switches can be used to measure rates. For example, tdescrete sensors can be used to measure the volume of fuel, water or chemical solution passing through a totalizing flow meter with a magnetically activated switch, or the speed of a rotating flywheel.

Sensors are an extremely important component of the control loop because they provide the basic data that drive an automatic control system. Understanding the operating principle of a sensor is very important. Sensors many times do not react directly to the variable being measured. For example, when a mercury thermometer is used to measure temperature, temperature is not being measured, rather, a change in volume due to a change in temperature is measured. Because there is a unique relationship between the volume and the temperature the instrument can be directly calibrated to provide temperature readings. The ideal sensor responds only to the "sensed" variable, without responding to any other change in the environment. It is important to understand that sensors always have a degree of inaccuracy associated with them and they may be affected by other parameters besides the "sensed" variable. The classical example is that of soil moisture measurement using electrical conductivity probes. The electrical signal produced by this sensor is closely related to soil moisture, but is greatly affected by temperature and dissolved salts (fertilizers, etc.) in the soil. Another important factor related to the sensor is its time response. A sensor must deliver a signal that reflects the state of the system within the frame of time required by the application. Using the soil moisture measurement example, the sensor must be able to "keep up" with the changes in soil moisture that are caused by evapotranspiration. Thus, proper selection of the sensors and understanding the principle of operation is critical to the success of a control system. Some of the variables that are often measured in computer based control systems are the following:

1. Flow rate,

2. Pressure,

3. Soil-moisture,

4. Air temperature,

5. Wind speed,

6. Solar radiation,

7. Relative humidity,

8. Total salts in irrigation water, and

9. pH of irrigation water.

A/D interface

Since computer systems work internally with numbers (digits), the electrical signals resulting from the sensors must be converted to digital data. This is done through specialized hardware referred to as the Analog-to-Digital (A/D) interface. Discrete signals resulting from switch closures and threshold measurements are converted to 0 and 1. Continuous electrical (analog) signals produced by the sensors signals are converted to a number related to the level of the sensed variable. The accuracy of the conversion is affected by the resolution of the conversion equipment. In general, the higher the resolution the better the accuracy. For, example if a pressure sensor produces a voltage signal ranging from 0 to 5 volts for a range of pressure of 10 atmospheres, an 8 bit resolution A/D board will be able to detect a change in voltage of about 5/255 volts which will results in measurable increments of 10/255 atmospheres. If the resolution of the A/D board was 12 bit, the board would be able to detect a change in voltage of about 5/4095 volts or a measurable increment of 10/4095 atmospheres.

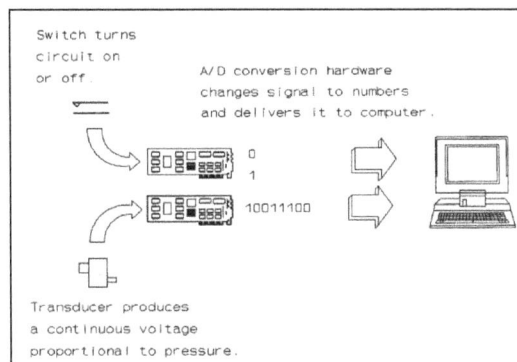

Computer System

The A/D conversion hardware is directly connected to the computer system. Given the current state of technology, the computer system may be a PC (personal computer), a minicomputer, or a specially designed machine that is solely dedicated to the control task. The type of machine depends on the type of application, and is greatly affected by factors such as environment characteristics, complexity of the controlled system, and the speed with which conversions need to take place (controlling a high speed extruder requires much more speed than a golf-course irrigation system). Many agricultural applications can be economically carried out using personal computers (PC), as is evident by the increasing number of system integrators and equipment manufactures that are marketing PC-based control systems. Also, many manufacturers of control equipment have designed and manufactured specialized computer control systems.

Control Interface

Using control software, decisions may be made to modify the controlled system. The actual changes are achieved by having devices within the system that will affect the controlled variables. These devices are controlled through actuators that respond to signals from the control interface. The devices may be of the nearly continuous or discrete types. For example, the extension of a robot arm of a citrus harvesting robot requires the use of a continuous signal from the computer, while a fan or a valve requires only an on/off (discrete) signal from the computer. In general, any device that can be powered electrically can be computer controlled.

Software Components

The software is used to implement procedures as they apply to the controlled system. These procedures are usually very elaborate, but in a well-engineered piece of software, they are transparent to the user. Because the user is more concerned with ease of use and performance of the system, good quality software has an interface that allows easy definition of the characteristics of the system to be controlled and simplifies the assignment of hardware resources. Performance is measured by how well the computer control system maintains the desired state.

Computer-based Controller Topologies

Centralized Computer

The simplest form in which a computer control system can be arranged is to use a single computer system that included the necessary support hardware and software to support data acquisition and control.

Satellite Systems

A centralized computer can be linked to other devices that have specialized purposes. One such type of device can be a datalogging system (a computer in itself) used to collect weather data. Also, in large complex systems, a central computer can be used to download instruction to intelligent controllers.

Communications between the central control computer can be implemented in a variety of ways:

a) Serial or parallel communication links.

b) Telephone link using serial communications.

c) Carrier wave using powerline modulation.

d) Communications bus.

Telephone and serial links can be hardwired or wireless.

Automatic Plant Irrigation System

Automatic irrigation system senses the moisture content of the soil and automatically switches the pump when the power is on. A proper usage of irrigation system is very important because the main reason is the shortage of land reserved water due to lack of rain, unplanned use of water as a result large amounts of water goes waste. For this reason, we use this automatic plant watering system, and this system is very useful in all climatic conditions.

The power supply consists of a step-down transformer, which steps down the voltage to 12VAC. By using a bridge rectifier this AC is converted to DC, then it is regulated to 5v using a voltage regulator which is used for the operation of the microcontroller.

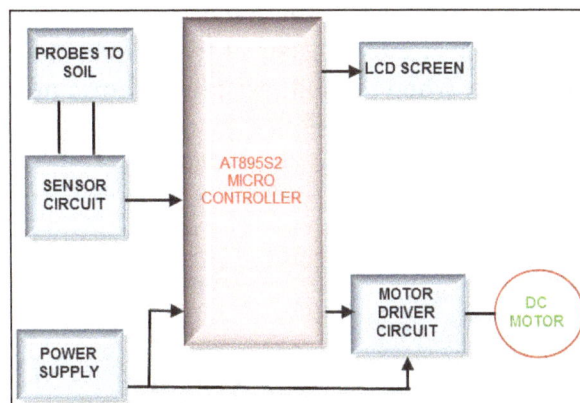

Block Diagram of Automatic Plant Irrigation System.

The block diagram of this automatic plant irrigation system comprises three main components namely a microcontroller, a motor-driver circuit and a sensor circuit. When the sensor circuit senses the condition of soil, it compares it with the reference voltage 5v. This process is done by a 555 timer.

When the soil condition is less than the reference voltage, i.e., 5v, then the soil is considered as dry and instantly the 555 timer sends the logic signal 1 to the microcontroller. The microcontroller then turns on the motor driver circuit and prompts the motor to pump water to the plants. When the soil condition is greater than the reference voltage, the soil becomes dry. Then the timer sends the logic signal 0 to the microcontroller, this turns off the motor driver circuit and prompts motor to pump water to the fields. Finally, the condition of the motor and soil are displayed in the LCD display.

Circuit Diagram of Plant Irrigation System

The main component used in this automatic plant irrigation system is 7404 Hex Inverter. The main function of the inverter output is proportional to input. It means, if the input of the inverter is low, then the output of the inverter will be high, and the inverter will give low output if the input is high. The Hex inverter 7404 IC includes six independent inverters and the range of operating voltage is around 4.75V to 5.5V, and the Supply voltage is 5V. They are used in many applications such as drivers, inverting buffers, etc. This IC is available in different packages like quad-flat package and dual-inline package. The pin configuration of the 7404 IC is shown below.

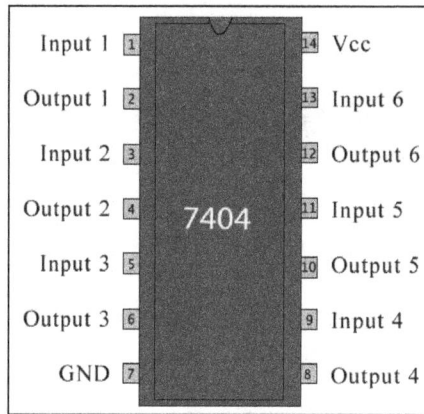

7404 IC Pin Configuration.

The circuit diagram of the plant-irrigation system is shown below. To make the circuit work and to water the pants, we use this simple logic: when the soil is dry, it has high resistance and when the soil is wet it has low resistance. This circuit consists of two probes that are placed into the earth. These probes perform the work only when the soil resistance is low and they cannot perform when the resistance of the soil is high.

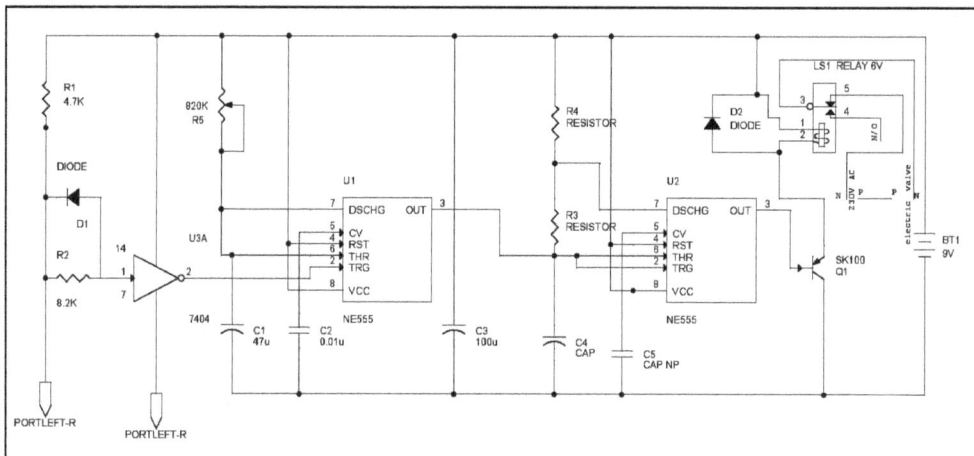

Plant Irrigation System Circuit Diagram.

To conduct the probes, the voltage supply is provided from a battery, which is connected to the circuit. When the soil becomes dry, it produces large voltage drop due to high resistance, and this is sensed by the hex inverter and makes the first NE555 timer. This timer is arranged as a monostable multivibrator with the help of an electrical signal.

When the first 555 timer is activated at pin2, it generates the output at pin3; and, this output is given to the input of the second timer. This second NE555 timer is configured with astable multivibrator and generates the output to make the relay which is connected to the electrically operated value through the SK100 transistor. The output of the second timer switches on the transistor that drives the relay. This relay is connected to the input of an electrical value and the output of the electrical value is given to the plants through the pipe.

When the relay is turned on, the valve opens and water through the pipes rushes to the crops. When the water content in the soil increases, the soil resistance gets decreases and the transmission of the probes gets starts to make the inverter stop the triggering of the first timer. Finally the valve which is connected to the relay is stopped.

The advantage of using an automatic plant irrigator is that it is a simple system capable of conserving water, improving growth, discouraging weeds, saving time, and controlling fungal diseases and adaptable to the conditions.

Micro Irrigation System

Micro irrigation is nothing but a slow and regular application of water and nutrients moving down drop-by-drop directly to the root zone of the plants through low-discharge emitters and plastic pipes. This irrigation system is today's need of the hour as the natural water resources which are gift to the mankind have become scarce, and that are now not unlimited and free forever. But, the world's water resources are now fast moving back on track. After one completes the study of inter relationship between crops, soil, water and climatic conditions, one will find this micro irrigation system as a suitable system capable of delivering exact quantity of water at the root zone of the plants.

Micro Irrigation System.

This system ensures that the plants do not endure from the strain or stress of less and over watering. The advantages of using this micro irrigation system are that for every drop of water used, we get more crop, better quality, early maturity, higher yield. Moreover, this system saves labor cost and water up to 70%. The working of this irrigation system covers over 40 crops spanning across 500 acres.

Automated Rain Barrel Drip System

Automatic Valves for Rain Barrels

Standard solenoid irrigation valves don't work well with a typical rain barrel. The standard solenoid valves used for most irrigation systems simply need more pressure than you have available from a typical gravity fed rain barrel. The higher pressure requirement for the valve is a function of the hydraulics that makes the valve operate. You either need more pressure or you need a different type of automatic valve. If you want to create more pressure you need to raise the height of the rain barrel. For every foot you raise the rain barrel you will create 0.433 PSI. The minimum operating pressure of most irrigation valves is at least 15 PSI, that means the barrel needs to be 34 feet above the height of the valve. That is simply not practical in most cases. Now you understand why those water towers you see in some communities are so tall.

Motorized Rain Barrel Valves

Mechanical motor-operated ball or butterfly type valves will open at any water pressure, unlike solenoid irrigation valves. This makes them a good solution for rain barrels irrigating by gravity alone.

One of the least expensive solutions is a combination timer and valve made for garden hoses. The Toro #53746 Battery Operated Hose End Timer is an example of this type of timer/valve. There are likely other brands available as well. This Hose End Timer uses a motorized ball valve to control the water flow. Most of the time the hose end timer gets the job done when used with a rain barrel but this is a low end market product and be aware that the quality is low. It may very well quit working after a year or two. On the other hand you can buy and replace a lot of these for the price of a full blown commercial quality motorized valve and timer unit like the ones they use on home floor heating systems.

A more expensive, but more reliable and longer lasting method, is to use a motorized ball valve made for home hydronic heating systems. Make sure you get a motorized ball valve, not a heat motor valve, unless you really want to use lots of power and take several minutes to open or close. Most of these hydronic valves are 24VAC and thus directly compatible with standard irrigation timers. So when using a motorized heating valve make sure the motor operates on 24VAC and a amperage your timer can handle. The hydronic valves come in a variety of voltages and amperages. Irrigation timers only work with low amperage 24VAC valves. To find these motorized valves do a search for "hydronic zone valve". Be sure to note the connection types for the valves, most are made to connect to PEX pipe or be soldered onto copper. You may have to install adapters to fit them to your irrigation system pipes or tubes.

Use Emitters that Work Well at Very Low Pressures

Drip Emitter and Tube Selection

Most people use drip irrigation with their rain barrels, (If you want to use sprinklers you will probably need a lot more water pressure, and therefore a larger pump.) The best emitters for the very low pressures in a rain barrel fed system are the most simple emitters, such as those commonly called a "flag emitter" or "take-apart emitter".

Another popular choice for emitters when using a rain barrel is the adjustable flow emitter/ bubbler. These use more water and are even less uniform than the Flag Emitters, but they are particularly good for watering pots of various sizes as you can adjust the flow needed for each pot.

Stay away from the higher cost emitters and those labeled as "pressure compensating" as they tend require higher pressures to operate efficiently.

Use 1/2" tube if you can and keep the drip tube lengths short. Smaller diameter tubes (especially 1/4" tube!) and longer tube lengths both restrict the water flow and lower the water uniformity between plants. To put it another way; if you use long, small tubes the plants closest to the rain barrel may drown from too much water, while the plants at the far end of the tube may not get any water at all.

Gravity fed drip systems from rain barrels are going to have less uniform water distribution. That's just the way it is, with minimal water pressure it is very hard hydraulically to maintain uniformity. To make the best out of a bad situation you use larger diameter tubes and keep the barrel as close as possible to the plants so the tubes are not too long. If that won't work for you, then the alternative is to use a small pump to create more water pressure. Most people just elect to be content with the low uniformity.

If you want to test the uniformity of your drip system it is very easy to do, simply build your drip system and attach it to your rain barrel. Then place a disposable plastic cup under each emitter and run the system for a few minutes. All the cups should have about the same amount of water in them. If the amount of water in the cups varies greatly then the uniformity is pretty bad. If the uniformity is bad enough that you think it will create uneven watering you can do a simple test to see if more pressure will help by hooking your drip system up to a garden hose. Be careful, the garden hose will provide more pressure than you need, so turn the valve on slowly and don't turn it on all the way. Empty out the cups and run the test with the cups again. Usually the higher pressure from the garden hose will result in more uniformity between the water in the cups.

Using a Pump for your Rain Barrel

The best way to automate a rain barrel irrigation system may be to not using a valve at all! Consider using a small pump placed on your rain barrel outlet hose. A pump is often the best solution as it may provide the added benefit of more water pressure. But it's not cheap to add a pump.

Selecting and Installing a Rain Barrel Pump

Make sure the pump is rated for enough flow to supply your emitters, and enough lift to get the water needed for your irrigation over the top of the barrel. Add the flow rate of all the emitters together to determine the flow rate needed for the pump. For example if you have 15 emitters that are rated at 1gph (gph means "gallons per hour") then the pump will need to supply at least 15 gph. If the barrel is 5 feet tall then the pump will need to lift the water 5 feet just to get it out of the barrel. But you need to do more than get the water out of the barrel. You need pressure to move the water *efficiently* through the tubes and push it out through the drip emitters. That requires another 45 feet of elevation. So add your rain barrel height to the elevation needed to power the drip system. 5 + 45 = 50 feet. So you want a pump with the capacity to move 15gph of water and lift it 50 feet.

Some pumps are rated using PSI (pounds per square inch of pressure) output value rather than feet of lift. A simple formula converts feet to PSI. Just multiply feet x 0.433 to get PSI. So a pump with a 50 feet of lift becomes 22 PSI. (50 feet * 0.433 = 22 PSI) So if the pump is rated in PSI it needs to produce 22 PSI.

If you can find one the right size, a submersible pump is the easiest and best method. Unfortunately most are made to be fountain pumps or sump pumps and they don't create enough water pressure to efficiently operate drip emitters. If you find one that will work for you, attach your irrigation hose to the pump, put the pump in the bottom of the barrel, and run the tube up over the top of the barrel. You will need a air vent at the high point on the tube near the top of the barrel (above the maximum water level) to prevent water from siphoning out of the barrel through the tube when the pump is not running. You can buy an air vent from any drip irrigation store. Or a very simple and cheap way to create an air vent is to add a drip emitter on the hose at the top of the barrel, so that the water from the emitter drips back into the barrel and is not wasted. When the pump turns off, this emitter will allow air to flow back into the tube and the air will stop the water from siphoning out.

If you don't use a submersible pump then the pump will be attached to an outlet at the bottom of the rain barrel. Make sure the pump is bolted or screwed down to a firm surface or it will jump all over the place when it runs. The tube from the pump outlet will need to be looped up above the top of the barrel and an air vent (or emitter as described above) installed at the high point to prevent all the water in the barrel from draining out through the pump when the pump is off.

Example of a pump: Surprisingly the best pumps for rainbarrels are not ones made for irrigation uses, but rather industrial pumps. For example the Little Giant 35-OM pump is made for high pressure applications like commercial carpet cleaners, but it produces good pressure at a low flow, a combination that is great for small drip systems. Here is the performance chart for this pump:

- 40 gph at 70 ft hd
- 60 gph at 65 ft hd
- 80 gph at 58 ft hd
- 100 gph at 54 ft hd
- 120 gph at 45 ft hd
- 140 gph at 30 ft hd.

Controlling the Rain Barrel Pump

The pump can be turned on and off by using a timer. A simple lamp or other household electricity timer will often work for an extreme low cost option, however lamp timers are pretty limited. Most timers of this type will only turn on and off the pump once a day, and do it every day. Most people don't need to water daily, so this could waste water. If you do use a simple timer make sure it is rated for a voltage and amperage that is equal to or higher than the input of your pump.

If you want to use a standard irrigation timer to control the pump you will need to buy a pump relay unit. Irrigation timers output 24 VAC, most pumps use 120 VAC. So the pump can't be connected directly to the irrigation timer. A relay is used to allow the pump to be turned on by the timer. Make sure the relay is rated for the correct voltage and amperage for your pump. Instructions for installing and wiring the pump relay should be provided with the pump relay.

Drip Irrigation System Plan

Five considerations for planning a drip irrigation system for large gardens or small market-grower operations.

1. Determine the quality of your water source for your drip irrigation. Factors may include such things as a pond water source that will require installation of a filtering system or an adjustment of the water pH, depending on your crops.

2. Familiarize yourself with the elevation of the plot/field you plan to irrigate, as it will determine how you size the system and adjust water flow. A 2.3-foot change in elevation, for instance, results in a gain of 1 pound of water pressure going downhill, or loss of 1 pound of pressure going uphill, requiring pressure compensation within the system on steep slopes. Topography, water flow rate and distance also will affect the size of pipes you'll need.

3. Will you automate your drip system? Depending on the complexity and size of the system, you may need to split watering times between different zones to water spaces incrementally, based on the output of your pump or the water needs of different crops. Automation ensures consistency in soil moisture and in flow, versus turning the water on and off at irregular intervals. The latter is important if you use the drip system to fertilize.

4. Think ahead. As plants mature, they require more water, which is especially important if you are irrigating perennial fruit crops. Build the system with the capacity to supply the optimum amount of water plants will need at maturity. If your irrigation water comes from the same well your home uses, the pressure tank may need to be upsized to reduce pump cycling and possible pump burnout. Or, you can irrigate at night when family water use is minimal.

5. Allow for expansion when installing your system. For systems one-half acre or larger, scaling for expansion up front will save money, as permanently installed pipes should be buried below the frost line and the cost of trenching in pipelines is costly. Size your pipes to accommodate future growth of your operation. Doubling the pipe diameter will quadruple the potential water flow rate.

Evaluation of Rotating Head Sprinklers and Operation of Sprinkler System

Evaluation of a Sprinkler Head

The sprinkler head is evaluated based on the water distribution pattern from the sprinkler nozzle, discharge of sprinkler nozzle, radius of throw, sprinkler rotation and precipitation or water application depth collected in a standard catch.

Site Conditions and Test Equipment

i) Sprinkler site: The sprinkler should be located in an area where the surface is smooth or where vegetative growth is less than 150 mm in height. The surface grade should not exceed 2 percent within the wetted area of sprinkler under test.

ii) Collector Description and Location: The collectors or catch cans used for any one test should be such that the water does not splash in or out. The type of collector should be identified and recorded on the data sheet. If an evaporation suppressant is used its type and method of application should be identified and recorded on the data sheet. The spacing of the collectors depends on the radious of throw of the sprinklers.

Table: Spacing of collectors.

Sprinkler radius of throw, m	Maximum collector spacing center to center, m
0.3-3	0.30
3-6	0.60
6-12	0.75
>12	1.50

iii) Sprinkler mounting: The sprinkler nozzle height above the nearest collector(s) for test purposes.

Table: Nozzle heights

SL. No.	Sprinkler type	Maximum nozzle height above collector (mm)
1	Riser mounted, rotating sprinkler of not more than 30 mm nominal inlet size	915
2	Riser mounted, rotating sprinkler of not less than 30 mm nominal inlet size	1830
3	Riser mounted, non-rotating sprinkler	460
4	Grade mounted sprinkler	Sprinkler lid level with the collector in the non-operating position
5	Hose end base mounted sprinkler	Bottom of sprinkler base to be level with the collector inlet.

iv) The sprinkler should remain vertical (within 20) throughout the duration of the test.

v) The position of all collectors should be maintained such that the entrance portion is level.

vi) The height of the top of any collector should be a maximum of 0.9 m above the ground.

Wind Measuring Equipment and Location for Outdoor Tests

The sprinkling pattern is influenced by wind; hence the mesiurment of wind velocity and direction are required to be known for sprinkler performance. Wind velocity should be measured with a rotating cup anemometer. The wind direction should be determined with a wind vane. Wind velocity sensing equipment should be located at a minimum height of 4.0 m. These equipments should be located outside the wetted area of the sprinkler and at a location that is representive of the wind conditions at the sprinkler location. The maximum distance of the sensor location should exceed 45 m from the wetted area of the sprinkler under test.

Measurements

i)   Sprinkler Pressure: The sprinkler base pressure should not vary more than ± 3 percent during the test period. Pressure should be measured with pressure measuring device accurate within ± 3 percent of the sprinkler test pressure and recorded in kPa. The pitot tube is the commonly used pressure measuring device for measuring the pressure at the nozzle of the sprinkler.

ii) Sprinkler Flow: The flow through the sprinkler should be measured to an accuracy of ± 3 percent of the sprinkler flow rate and recorded in m3/h. Data rates up to 95 m3/h. Rates than 95 m3/h should be listed to at least the nearest 0.2 m3/h. Data rates up to 95 m3/h. Rates than 95 m3/h should be listed to at least the nearest 0.2 m3/h. The flow rate can be measured by connecting the tube to the nozzle and measuring the volume of water collected in a water tank for a specified time.

iii) Sprinkler Radius of Throw

   a)   The radius for rotating sprinklers should be defined as the distance measured from the sprinkler centerline to the farthest point at which the sprinkler deposits water at the minimum rate of 0.25 mm/h over the inlet surface area of the collector.

   b)   The radius for non-rotating sprinklers should be defined as the farthest distance measured from the sprinkler centerline to the point at which the sprinkler deposits water at the minimum rate of 0.25 mm/h typically measured at any arc of coverage except at the arc extremes of part circle sprinklers.

   c)   The radius of throw for both full and part circle sprinklers should be reported to the nearest 0.3 m.

iv) Sprinkler Rotation: The sprinkler rotation speed should be measured only while the sprinkler is rotating from its own drive mechanisms and should be recorded.

v)   Collector Readings: The amount of water in each collector should be accurately determined and recorded showing the location of the collectors relative to the sprinkler. For multi-leg tests, the reading for each leg should be recorded independently.

vi) Test Records and Data Recording: The data outlined in this topic should be recorded on appropriate forms. Supplemental data describing the conduct of the test may be included on the form.

Moisture Distribution Pattern and Uniformity of Coverage

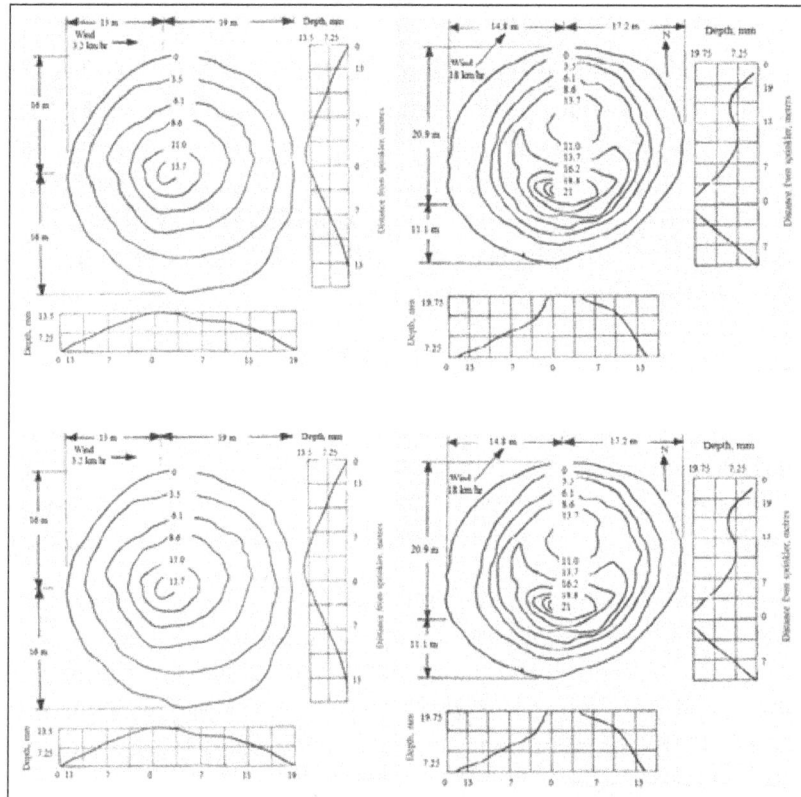

Moisture distribution pattern from a rotating head sprinkler under
favourable & windy conditions of pressure and wind.

The application efficiency of sprinkler depends upon the degree of uniformly of water application. The basic objective of sprinkler irrigation is to apply uniform depth of water at a given application rate. The uniformly of water application depends upon the water spray distribution characteristics of sprinkler nozzle and sprinkler spacing. The spray distribution characteristics change with nozzle size and operating pressure. The drops are larger and the water from the nozzle falls in a ring away from the sprinkler at lower pressures. For higher pressures, the water from the nozzle breaks up into very fine drops and falls close to the sprinkler. External factors such as wind also distorts the application pattern. Higher the wind velocity, greater the distortion and this factor should be considered when selecting the sprinkler spacing under windy conditions. This distribution pattern from sprinklers for favorable wind conditions and optimum pressure is shown in figure. It can be seen that the depth of water applied surrounding the sprinkler decreases with increase in the distance from the sprinkler. Similar pattern of the water in the soil can be observed in figure. The figure clearly shows that the pattern of moisture distribution is not uniform with the single sprinkler. Therefore to obtain the uniformity in water application, it is necessary that the moisture distribution pattern of the adjacent sprinklers be overlapped properly. Figure shows the water distribution pattern of overlapped sprinklers. The wetted circles formed by adjacent sprinklers are overlapped so as to add water to areas of the adjoining sprinklers for obtaining the depth of water application. The aggregate depth of distribution obtained by overlapping thus becomes nearly uniform as shown in figure.

The figure also shows the moisture distribution pattern of a rotating head sprinkler under windy conditions and corresponding moisture distribution in soil.

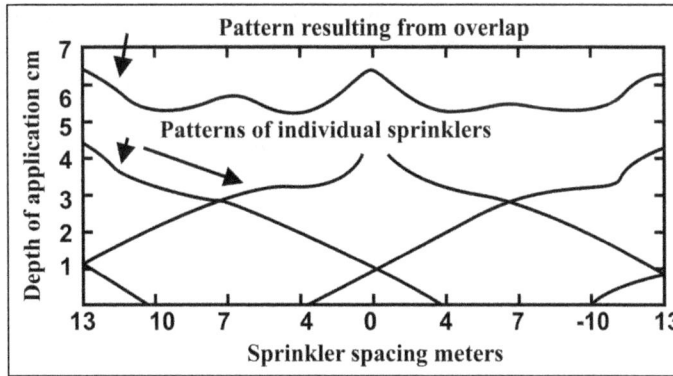

Water distribution pattern from overlapped sprinklers.

Uniformity of Coverage

Measuring Distributions: The distribution of sprinkler systems can be evaluated by measuring the patterns of individual sprinklers, and then by combining, or by sampling directly, ASAE Recommendations S330 describes procedures for measuring the distribution of a single sprinkler including the format for presenting the data. According to ASAE measurements the test site should be nearly level and the minimum clean distance upwind should be positioned at least 60 cm above the collectors, and 90cm above the surface. Preferably, the sprinkler should be located in the center of the grid of the four adjacent central collectors. A minimum of 80 collectors should receive water during a test. Wind direction and total wind movement at the 4-m height should be recorded for interpretation of the data. The distribution of a typical sprinkler system can be evaluated with a grid of catch cans or collectors. The grid should be located over a length equal to at least the sprinkler spacing on the lateral and over a width at least equal to lateral spacing. For linear moving systems, one or two lines of measuring devices perpendicular to the travel path may be more practical and meaningful than a grid system.

Uniformity Coefficients (C_u): It is a measurable index of the degree of uniformity obtainable for any size sprinkler operating under given conditions. This uniformity coefficient is affected by the pressure nozzle size relations, sprinkler spacing and by wind conditions. The coefficient is computed from field observations of the depths of water collected in catch cans or collectors placed at regular intervals within a sprinkled area as per procedure described in preceding sections. It is expressed by the equation developed by Christiansen:

$$C_u = 100\left(1.0 - \frac{\sum X}{mn}\right)$$

in which,

C_u = coefficient of uniformity

m = average value of all observations (average application rate), mm

n = total number of observation points

X = numerical deviation of individual observation s from the average application rate, mm.

A uniformity coefficient of 100 per cent (obtained with overlapping sprinklers) is indicative of absolutely uniform application, whereas the water application is less uniform with a lower value of coefficient. A uniformity coefficient of 85 per cent or more is considered to be satisfactory.

Pattern Efficiency: The pattern efficiency (also known as distribution efficiency)is calculated with the total depths of water collected at each of the catch cans placed at the grid points. The minimum depth is calculated considering average of the lowest one fourth of the depths collected in catch cans used in a particular test. Pattern efficiency is given by,

$$P_e = \frac{Min.\,depth}{Average\,depth} \times 100$$

The pattern efficiency is useful in calculating the average depth to be applied for a certain minimum depth. The pattern efficiency is influenced by the wind conditions.

The application efficiency is given by,

$$Application\ efficiency = \frac{Min.\,rate\,caught}{Average\,rate\,applied}$$

Operation and Maintenance

The operation mode for a solid-set or permanent sprinkler system depends upon the design and use of the system, available labor, water supply, and available capital. Either system can be designed on the lateral or area (block) design method. With the lateral design method, individual laterals are controlled by valves and each lateral may be operated as desired. Normally, more than one lateral is operated simultaneously, but the operating laterals usually are widely separated in the field. The lateral design method minimizes the main or supply line pipe size, but it increases the number of valves required and also the time to open and close valves when a manually operated valve system is used. With the area (or block) design method, a contiguous portion of the field is irrigated at one time. Usually a sub-main is installed to supply water to that portion of the field.

For frost and snow protection, the entire system may be operated at one time. Depending upon the crop being protected, the application rate will be 2 to 5 mm (0.08 to 0.18 in.) per hour. Both undertree and overtree systems are used; however with saline water only undertree systems should be used. Single nozzle, medium pressure sprinklers should be used for frost and snow protection. For crop cooling and blossom delay, the entire system may be sequenced in alternate on-off modes as one portion of the system may be operated at a time and the operation can be switched to another portion of the system, Sequencing is best accomplished with electric controllers and automatic valves. If the system is being used strictly for irrigation, only a portion of the system is normally operated at one time. Where several hours are required for irrigation, control may be manual or automatic.

A sprinkler system like any other farm equipment needs maintenance to keep it operating at peak efficiency. Parts of the system subject to wear are the rotating sprinkler heads, the pumping set, the couplers and the pipeline. General principle regarding the maintenance of the pipes and fittings and sprinkler heads are given below:

- Pipes and Fittings: The pipes and fittings require virtually no maintenance but attention must be given to the following procedures:

    (a) Occasionally clean any dirt or sand out of the groove in the coupler in which the rubber sealing ring fits. Any accumulation of dirt or sand will affect the performance of the rubber sealing ring.

    (b) Keep all nuts and bolt tight.

    (c) Do not lay pipes on new damp concrete or on piles of fertilizer. Do not lay fertilizer sacks on the pipe.

The pipes are automatically emptied and ready to be moved. When the pump is first started and before the pressure has built up in the system the seals may give a little leakage. With full pressure in the system the couplers and fittings will be effectively leak free. If however there is a leakage check the following:

(a) There is no accumulation of dirt or sand in the groove in the coupler in which the sealing ring fits. Clean out any dirt or sand and refit the sealing ring.

(b) The end of the pipe going inside the coupler is smooth clean and not distorted.

(c) In the case of fittings such as bends, tees and reducers ensure that the fitting has been properly connected into the coupler.

- Sprinkler Heads: The sprinkler heads should be given the following attention.

    (a) When moving the sprinkler lines make sure that the sprinklers are not damaged or pushed into the soil.

    (b) Do not apply oil, grease or any lubricant to the sprinklers. They are water lubricated and using oil, grease or any other lubricant may stop them from working.

    (c) Sprinkler usually have a sealed bearing and at the bottom of the bearing there are washers. Usually it is the washers that wear and not the more expensive metal parts. Check the washers for wear once a season or every six months this is especially important where water is sandy. Replace the washers if worn.

    (d) After several seasons operation the swing arm spring may need tightening. This is done by pulling out the spring end at the top and rebending it. This will increase the spring tension. In general check all equipment at the end of the season and make any repairs and adjustment and order the spare parts immediately so that the equipment is in perfect condition to start in the next season.

- Storage: The following points are to be observed while storing the sprinkler equipment during the off season:

    (a) Remove the sprinklers and store in a cool dry place.

(b) Remove the rubber sealing rings from the couplers and fittings and store them in a cool, dark place.

(c) The pipes can be stored outdoors in which cases they should be placed in racks with one end higher than the other. Do not store pipes along with fertilizer.

(d) Disconnect the suction and delivery pipe work from the pump and pour in a small quantity of medium grade oil. Rotate the pump for a few minutes. Blank off the suction and delivery branches. This will prevent the pump from rusting. Grease the shaft.

(e) Protect the electric motor from the ingress of dust, dampness and rodents.

Common Troubles and Remedies in Operation of Sprinkler System

The following are the general guidelines to identify and remove the common troubles in the sprinkler systems:

Pump does not Prime or Delivers Water:

(a) The pump suction lift should be checked, if it is within the limits. If not lower the pump closer to the water.

(b) Check the suction pipeline and all connections for air leaks. All connections and flanges should be air tight.

(c) Check that the strainer on the foot valve is not blocked.

(d) Check that the flap in the foot valve in free to open fully.

(e) Check the pump glands gently. If necessary repack the glands using a thick grease to seal the gland satisfactorily.

(f) Check that the gate valve on the delivery pipe is fully closed during priming and opens fully when the pump is running.

(g) Check that the direction of rotation of the pump is correct.

Sprinklers do not Turn

(a) Check pressure.

(b) Check that the nozzle is not blocked. Preferably unscrew the nozzle or use a small soft piece of wood to clear the blockage. Do not use a piece of wire or metal as this may damage the nozzle.

(c) Check that the sprinkler can usually be pushed down towards the riser pipes so that the water pressure flushes out the bearing. If the bearing is still stiff dismantle and then clean it. Do not use oil, grease or any lubricant.

(d) Check that the condition of washers at the bottom of the bearing and replace then if worn or damaged.

(e) Check that the swing arm moves freely and that the spoon which moves into the water stream is not bent by comparing it with a sprinkler which is operating correctly. If it is bent then very carefully bend it back into position.

(f) Adjust the swing arm spring tension. Usually it should not be necessary to pull up the spring by more than about 6mm.

Leakage from Coupler or Fittings

The sealing rings in the couplers and fittings are usually designed to drain the water from the pipes when the pressure is turned off. This ensures that the pipes are automatically emptied and ready to be moved. When the pump is first started and before the pressure has built up in the system the seals may give a little leakage. With full pressure in the system the couplers and fittings will be effectively leak-free. If, however, there is a leakage, check the following:

(a) There is no accumulation of dirt or sand in the groove of the coupler in which the sealing ring fits. Clean out any dirt or sand and refit the sealing ring.

(b) The end of the pipe going inside the coupler is smooth, clean and not distorted.

(c) In the case of fittings such as bends, tees and reducers ensure that the fitting has been properly connected into the coupler.

Design and Operation of Underground Pipeline System

The design of underground pipe line system requires information on land topography, location of water source and water discharge. Pump stands must be of high elevation to allow sufficient operating head for the pipeline. However, stands higher than necessary may permits high heads of water to build up, leading to excessive line pressures. The working pressures in the pipeline are kept within one-fourth the internal bursting pressures of the pipe. When it is necessary to design pipelines with higher heads, reinforced concrete pressure pipes are used. The sizes of the outlets are selected to suit the flow required at diversion points. The PVC and HDPE are also used for water distribution at low and moderate pressure. The components of the systems such as pipeline size and height of Pump stands and control stands must be designed so as to obtain a balanced water distribution and provide trouble free operation.

The height of water in the pump stands is estimated as follows:

Depth of water in pump stand (H_{ps}) (m) = Reduced level at height point (m) + losses in the pipe line Reduced level at pump stand (m). A free board of 0.5m of water head is added to get the height pump stand.

Losses in pipe line are head loss due to friction and also known as major loss. Various equations such as Darcy-Weisbach, Hazen Williams and Scobey have been proposed to determine head loss due to friction. The Darcy's Weisbach equation is scientifically based and applicable to both laminar and turbulent flows.

The Darcy-Weisbach equation is,

$$H_f = f \frac{L}{D} \frac{V^2}{2g}$$

where,

H$_f$ = head loss due to friction, (m)

f = friction factor,

L = length of the pipe(m)

d = Inside diameter of the pipe (m)

V = mean velocity of flow (m s^{-1})

g = acceleration due to gravity (m s^{-2})

The friction factor f is function of Reynolds number (Re) and relative roughness. For laminar flow (Re , the friction factor is):

$$f = \frac{64}{R_e}$$

Where,

$$R_e = \frac{VD}{v}$$

V = velocity, m s^{-1}

D = diameter of pipe, m

v = Kinematic viscosity, m^2 s^{-1}

For Re between 2000 and 100,000 (Turbulent flow),

f = 0.32 Re -0.25

For Re > 100,000 (Fully turbulent flow),

$$f = 0.80 + 2.0 \left(\frac{Re}{\sqrt{f}} \right)$$

For most commercial materials, the friction factor is represented by the semi- empirical equation:

$$\frac{1}{\sqrt{f}} = 1.14 - 2\log\left(\frac{e}{D} + \frac{9.35}{Ref} \right).$$

Minor Losses

Head losses in underground pipeline are also caused by inlets, bends, gate valves, outlets (rivers, valves etc.) and other appliances such as fittings expansions and constructions due to entry and exit losses and abrupt and gradual changes in velocity. These losses are, referred to as minor losses.

These losses are given by equation $K_1 \dfrac{V^2}{2g} + K_2 \dfrac{V^2}{2g} + K_3 \dfrac{V^2}{2g} + \ldots\ldots K_n \dfrac{V^2}{2g}$ Each term in the equation represents the head loss due exit, entry or fitting or connections in the pipeline:

$$K_1 \frac{V^2}{2g} + K_2 \frac{V^2}{2g} + K_3 \frac{V^2}{2g} + \ldots\ldots K_n \frac{V^2}{2g}$$

Where,

K_1, K_2,... K_n are coefficients for each item where minor head loss exists.

These minor loss coefficients can be obtained from Michael.

The values of coefficient as 0.5 for pipe flush with wall, 0.1 for bell entrance and 1.0 for bends. are used. If the gated pipes are used, then pressure required to operate these pipes are included.

Diameter of Pipe Line

The diameter of pipe line is computed considering the head loss due to friction in pipe line (Equation $H_f = f \dfrac{L}{D} \dfrac{V^2}{2g}$) and discharge. Too small, diameter will increase the pumping cost due to increased frictional head losses and too large pipe diameter will add to the system cost. The material and size of pipe are selected considering the hydraulically efficiently and pumping cost.

Laying Out of Underground Pipeline System

Preparation of contour map is essential requirement to lay and construct underground pipeline for a command area. The map should depict North direction and important features locating revenue division and sub divisions. The map should show field boundaries, streams, rivers, tanks, earth embankments, roads, wells and village boundary and other major features of the area which will come under the command of the tubewell and the area immediately adjoining it. The alignment of the buried pipeline water distribution system and the location of the valves should be planned based on the inspection of the field contours and the various features on the ground. The alignment of the earthen field channels and essential field structures like inverted siphons are decided based on the ground elevation. Profiles along each pipe and channel route are surveyed and plans are prepared showing depth, gradient and earth work along their length and the location of the structures. These include inlet, water control and diversion structures, air release vents and end plugs.

After laying of PVC pipes and backfilling of the trenches and construction of the outlet structure, the water is allowed to pass through the pipeline. All air in the pipe should be allowed to escape through the pressure release pipe. For 3 to 4 days, all the outlets are kept open for 3 to 4 hours, with the pump in operation, in order to check if there are any leakages in the pipe or in any of the outlets. If leakage is noticed, repairs are done after draining the entire pipeline system.

Operation and Maintenance of Underground Pipeline Systems

The underground pipeline may fail due to:

1. Lack of inspection or maintenance,

2. Improper construction,

3. Improper design and

4. Wrong manufacturing processes and poor quality materials used.

The underground pipelines operate without trouble when it is properly designed and correctly installed. Inadequate procedures in design and installation and unforeseen situations give rise to the following troubles.

1. Development of longitudinal cracks in the pipe, usually at the top or both at top and bottom.

2. Telescoping of sections.

3. Pushing of the pipe into the stands.

4. Development of circumferential cracks.

5. Surging or intermittent flow of water.

Leak Testing and Repair

All buried low pressure irrigation pipelines should be tested for leaks before the trench is filled. The pipeline should be filled with water and slowly brought up to operating pressure with all turnouts closed. Any length of pipe section or joints showing leakage should be replaced and the line retested. The water should remain in pipelines throughout the backfilling of trenches, because the internal pressure helps to prevent pipe deformation from soil loading and equipment crossings. Underground pipelines should be inspected for leakage at least once a year. Leaks may be spotted from wet soil areas above the line that are otherwise unexplained. Small leaks in concrete pipeline can be repaired by carefully cleaning the pipe exterior surrounding the leak, then applying a patch of cement mortar grout. For larger leaks, one or more pipe sections may have to be replaced. Longevity of concrete pipelines can be increased by capping all opening during cold winter months to prevent air circulation. Small leaks in plastic pipe, except at the joints, can sometimes be repaired by pressing a gasket-like material tightly against the pipe wall around the leak and clamping it with a saddle. Where water is supplied from a canal to portable surface pipe, sediment often accumulates in the pipe. This sediment should be flushed out before the pipe is moved. Otherwise, the pipe will be too heavy to be moved by hand and may be damaged if it is moved mechanically. Buried plastic pipelines can be expected to have a usable life of about 15 years, if well maintained. The annual cost of maintenance can be estimated as approximately 1% of the installation cost.

References

- Irrigation-scheduling-for-regulated-deficit-irrigation, irrigation, soil-and-water, farm-management, agriculture: vic.gov.au, Retrieved 21 March, 2019

- Irrigation-Controller-Introduction: allianceforwaterefficiency.org, Retrieved 23 July, 2019

- Automatic-plant-irrigation-system-circuit-and-its-working: edgefx.in, Retrieved 3 February, 2019

- Automatic-rain-barrel-irrigation: irrigationtutorials.com, Retrieved 12 April, 2019

- Tips-for-planning-a-drip-irrigation-system, home-garden: myfarmlife.com, Retrieved 11 January, 2019

Chapter 4

Land Grading and Survey in Irrigation

Reshaping the land surface to planned grades for the purpose of irrigation and subsequent drainage is known as land grading. Leveling the land is an important aspect of irrigation since it ensures that the depths and discharge variations over the field are relatively uniform. The diverse applications of land grading, surveying and leveling have been thoroughly discussed in this chapter.

Land Evaluation

A fuller use of land and water resources by the development of irrigation facilities could lead to substantial increases in food production in many parts of the world. The process whereby the suitability of land for specific uses such as irrigated agriculture is assessed is called land evaluation.

Land evaluation provides information and recommendations for deciding 'Which crops to grow where' and related questions. Land evaluation is the selection of suitable land, and suitable cropping, irrigation and management alternatives that are physically and financially practicable and economically viable. The main product of land evaluation investigations is a land classification that indicates the suitability of various kinds of land for specific land uses, usually depicted on maps with accompanying reports.

Table: Structure of the Suitability Classification.

ORDER	CATEGORIES CLASS	SUBCLASS
S Suitable	S1	
	S2	S2t
		S2d
		S2td
		etc.
	S3	
N Not suitable	N1	N1y
		N1z
		etc
	N2	

Legend

- S1 Highly Suitable,

- S2 Moderately Suitable,

- S3 Marginally Suitable,

- N1 Marginally Not Suitable,

- N2 Permanently Not Suitable.

Lower case letters in a Subclass indicate the nature of a requirement of limitation (e.g. t and d for topography and drainage). Land suitability units (subdivisions of Subclasses) may also be used to indicate minor differences in management.

From Project Identification to Project Implementation

In the early stages of land resources investigations, land evaluation studies indicate in a preliminary way, the suitability of land for alternative crops and irrigation methods and the land improvements that may be worthwhile. With further field studies, projects are identified and a plan of irrigation development is worked out. Individual projects are ranked in order of priority. The priority projects are planned in more detail and each project plan is progressively refined. The proposed crops, methods of irrigation, inputs and land improvements are progressively adjusted until a satisfactory project plan is produced.

Various criteria are used to decide whether a project plan is satisfactory. Apart from social and political objectives, which in practice are often paramount, a satisfactory plan is one that leaves the farmers, the community and the national economy better off. In other words, it results in the largest practicable increment in net benefits in an economic comparison of 'without project' and 'with project' situations. Such a plan will generally utilize limited resources of land, water or inputs for the most productive use. A satisfactory plan is one which is practicable and likely to work out under actual field conditions, not necessarily the most economically attractive on paper.

Land evaluation reports, maps and data continue to be useful after the planning stage during design and implementation, and for monitoring the project.' The detailed design of engineering works may depend on information collected earlier during the evaluation study. During the implementation and later management of the irrigation project, the land evaluation study provides a basis for monitoring changes in physical, social and economic conditions. In response to such changes, the recommendations may need modification and updating from time to time.

Currently, the rehabilitation of existing irrigation projects is an important aspect of land evaluation work. This highlights the need for thorough evaluation of land and water resources in the preparation of irrigation projects from the start; obviating the need for later rehabilitation.

Levels of Intensity of Investigations

The study of land and water resources and the production of irrigation development proposals may be conducted at national level, at the level of a river basin or hydrogeological basin, at project development level, or at village, farm or field level. The types of studies undertaken at these different levels are indicated in table.

Table: Levels of Intensity of Investigations.

Levels	Types of study	Types of survey
National Basin	Project identification	Reconnaissance
Project	Pre-feasibility	Semi-detailed
	Feasibility	Detailed
Village, farm or field	Detailed design	Very detailed

Project identification at a national and basin level leads to a need for prefeasibility and feasibility studies. These are followed by detailed design studies of water supply systems and field layouts. These various studies are served by different scales of survey.

At the national level, investigations are required to provide a Master Plan for land and water resources development including an assessment of the priorities accorded to respective regions and areas within the country. At the level of individual river basins or hydrogeological basins, investigations provide the basis for water development, water control for different uses and for land use planning (e.g. catchment projection, flood zoning, potential areas for irrigation, reclamation of delta and swamp and tidal zones, etc.). At the irrigation development project level, a plan is formulated for investment in irrigation, drainage and flood protection. At the village, farm or field level, investigations provide information for farm water management and improvements or rehabilitation.

Reconnaissance surveys on a small scale i.e. 1:100 000 to 1:250 000 are useful for broad resource inventory, the identification of promising areas for development, and to provide a basis for more detailed study. Mapping units are usually compound and provide only estimates of the proportions of the conditions for the various land suitability categories. The 'land system' method of survey is often used and it may suffice to broadly distinguish land which is promising for specific kinds of irrigated agriculture from land which is not. Economic studies at this, stage broadly indicate levels of production and income.

Semi-detailed surveys in pre-feasibility and feasibility studies are typically at scales from 1:25 000 to 1:50 000. Soil mapping units consist of a mixture of homogenous units (soil series) and compound units (e.g. soil associations). With sufficiently intense sampling, such surveys can be used for planning some developments up to the design stage.

Table: Survey intensity, mapping scale and kind of maps.

Kind of survey	Range of scales	
Very High Intensity (very detailed)	Larger than 1:10 000	Soil maps showing special features or phases of soil series and occasionally soil complexes; detailed topographic maps with spot heights; cadastral maps; groundwater maps; present crops and vegetation etc.
High Intensity (detailed)	1:10 000 to 1:25 000	Soil maps showing phases of soil series and soil complexes; detailed topographic maps, groundwater maps, present land use, etc.
Medium Intensity (semi-detailed)	1:25 000 to 1:100 000	Soil maps showing series or associations of series; land system maps, physiographic units, topographic contour maps, present land use maps, etc.

Low Intensity (recon-naissance)	1:100 000 to 1:250 000	Soil maps with associations and phases of Great Groups or Subgroups; land system maps, physiographic or contour maps, present land use, climatic zones, etc.
Exploratory	1:250 000 to 1:1 000 000	Land units of various kinds.
Syntheses	Smaller than 1:1 000 000	Climatic maps, soil taxonomic maps, vegetation and land use, physiographic and geomorphological maps, agro-ecological zones, etc.

Detailed surveys may be required separately for soils and topography. Soil surveys, typically at scales of 1:10 000 to 1:25 000, with soils series and phases as the main soil mapping units, are used for project planning and implementation and for some surveys at village or catchment level, including layout of farms and irrigation systems. If topography is an important consideration in delineating land to be brought under command by gravity irrigation, a more intensive survey (e.g. at 1:5 000) may be required for land levelling and engineering applications.

Very detailed surveys, at scales of 1:5 000 or larger, are necessary where small contour intervals must be mapped in order to determine slope classes, or align irrigation and drainage channels.

Planning a Land Evaluation Investigation

Land evaluation investigations may be carried out by a government department or private company with or without external help. Large irrigation projects often involve a client, a funding agency and a consultant organization. Prior to field work, initial discussions will take place to decide the objectives of the evaluation, and the data and assumptions on which it is to be based. The extent and boundaries of the area to be evaluated, and the kinds of land use or irrigation systems may either be prescribed in the terms of reference, or may be part of the evaluation. Appropriate physical or economic measures of suitability must be decided. The intensity and scales of the required surveys, and the phasing of activities should be agreed prior to the start of field work. The administrative, logistical and financial implications of the work being undertaken should also be agreed.

During initial discussions the requirements for reports and maps at various stages during the study should be decided. The regular production of progress reports and maps is a feature of all efficiently organized irrigation development investigations. These are essential as a basis for making major policy decisions at crucial stages of the study. It is also customary to produce interim reports in order to facilitate discussions and amendments before producing the final reports and maps.

Surveying for Construction of Irrigation Projects

First step in construction of an irrigation project like dams, barrage or weir requires surveying of whole area. Surveying for an irrigation project is done to understand if the dams or other irrigation construction is required or not.

The area should benefit to its larger extent when the completion of irrigation structure is done. So, survey is needed to conclude this.

Surveying for Construction of Irrigation Projects

The steps involved in surveying to build an irrigation structure are:

1. Examine the water availability.

2. Examine the topography.

3. Selection of site.

4. River gauging.

5. Marking of CCM.

6. Marking of tentative alignment.

7. Reconnaissance survey.

8. Preliminary survey.

9. Final location survey.

10. Final Survey Report.

Examine Availability of Water

To construct an irrigation project whether it is a dam or weir or barrage, first and foremost observation should be the presence of water and its availability.

The availability may be of different types, but proper examination is required before construction.

Some important observations are:

1. If there is any river flowing in that area, we should know the type of river whether it is perennial or inundation type. If it is perennial, then the water is available throughout the year. If it is Inundation River, then study its previous yearly discharges.

2. The river should meet the requirement of water in that area.

3. Suitable site be available to construct an irrigation project.

Examine Topography

After investigating the water availability, topography map of the area is studied. This study is more useful when marking the tentative alignment for irrigation project. The behavior of agricultural lands are examined in this stage.

Selection of Construction Site

When plenty of water or major source of water is available then the location to construct an irrigation project is selected. The project may be dam or barrage or weir.

The selection of site is done by considering the following points:

- The soil survey is conducted by boring and pile testing to know about the foundation depth required.

- Sufficient basin area should be available and the capacity must fulfill the required demand.

- The site should be easily accessible. Materials and labor should be readily available.

- The allowable bed slope should be maintained as far as possible in the canal.

- The structure should not submerge valuable lands and areas.

River Gauging

River gauging is measuring of water discharge at point. The point in this case where river gauging is conducted is the site selected for project.

After river gauging following details are obtained:

- The discharge of river is calculated on daily basis and the yearly discharge records are studied.

- The HFL (high flood level) and LWL (lowest water level) are recorded based on the old observations.

- To find out the possible silting of river bed, silt analysis is conducted, and manorial value of fine silt is recorded.

Marking of CCM

CCM is the cultivable command area which is mainly fit for cultivation of crops. The area under this category should be marked on the topographic map. So, the construction should not disturb or damage this area and required demand discharge can also be known.

Marking of Tentative Alignment

After the selection of site for irrigation structure, it is time to select the tentative alignments for canals or branch canals. These alignments should be marked in topographical and contour maps.

Marking should be done by following considerations:

- The alignment marked should cover the whole area when it is cut into canal.
- The alignment should minimize the earth filling and cutting costs.
- It should not pass through valuable agricultural lands, religious places etc.
- It should cross rivers, roads, rails etc. perpendicularly.

Reconnaissance Survey of Irrigation Projects

After marking the tentative alignments, then reconnaissance survey is conducted for all the alignments. This survey provides the following details:

- Nature of ground slope along the alignment.
- Magnetic bearings of lines of the traverse along the alignment are recorded.
- Alignments passing through religious places, valuable lands are eliminated. If they are unavoidable, they are marked as special areas and suitable compensation is provided.
- Nature of ground up to a distance of 100m on both sides of alignment are noted.
- Alignments cutting the crossings perpendicularly are noted.
- Distances are measured by pacing.
- Past records of floods in that area are noted.
- Suitable cross drainage works should be noted.
- If there is any river across the alignment, the alignment should cut the river across its shortest width.

Preliminary Survey of Irrigation Projects

After completion of reconnaissance survey, a good alignment is selected, and they are allowed to conduct preliminary survey.

Following steps are involved in this survey:

- Pillars are constructed on both banks of river and they represent the center line of irrigation project.

- Similarly pillars are constructed to mark the center lines of head works for both bank canals.

- Boring is done along the center line of irrigation structure to determine the depth of foundation.

- A permanent benchmark is created and its value is noted by connecting it to the nearby GTS benchmark by fly leveling.

- Plane table survey or prismatic survey is conducted on the both sides of alignments up to 100m and route survey map is prepared.

- Longitudinal leveling is conducted with an interval of 20 m.

- Cross leveling is conducted with an interval of 100 m.

- Permanent bench marks are arranged with some interval gap along the alignment.

- Water table level is studied on both sides of alignment covering up to 200 m.

- Soil along the alignment is surveyed.

- The details of road and railway crossings are noted to design cross drainage works.

- At river crossings, the cross section details of river are noted. The cross sections of river are taken on both upstream and downstream sides with covering of 500 m distance.

- Drawings are prepared for all the maps and cross sections.

- Estimate sheets are prepared for earth works, compensations for lands etc.

Final Location Survey

After preliminary survey, most economical alignment among all is selected and final location survey is conducted. In this final stage following steps are involved:

- Center line of final alignment is marked with pillars and pegs.

- The width of alignment is also marked by the pillars.

- Similarly for the branch canal also, pillars are marked.

- Final Eligible properties for compensation are recorded.

Final Survey Report of Irrigation Project

It is the last stage of the whole process and in this stage a report should be prepared with the details of final alignment.

This report is submitted to higher authorities to get an approval for an irrigation project:

- Introduction,

- Justification and necessity of project,

- Justification for the selection of final alignment,

- Detailed estimate sheets for earth works, compensation, head work, etc.

- Detailed specification for project,

- Benefit of project,

- Recommendation of the project,

- And following maps are to be submitted along with the above,

- General map of area through which canal passes,

- Route survey map,

- Longitudinal section of alignment,

- Cross section of alignment,

- Contour map of alignment,

- Drawings of dam, head works, cross drainage works etc.

Land Grading for Irrigation

Land grading is reshaping the surface of land to planned grades for irrigation and subsequent drainage. Land grading permits uniform and efficient application of irrigation water without excessive erosion and at the same time provides for adequate surface drainage. A plane surface (uniform row and cross slopes) is easiest to manage and maintain.

All lands to be graded for irrigation should be suitable for use as irrigated land and for the proposed methods of water application. Water supplies and the delivery system should be sufficient to make irrigation practical for the crops to be grown and the irrigation water application methods to be used.

Design Criteria

Soils should be deep enough so that, after the needed grading work is done, an adequate, usable root zone remains over most of the field that will permit satisfactory crop production with proper conservation measures. Limited areas with shallower soils may be graded to provide adequate drainage, irrigation grades or a better field arrangement.

All grading work for drainage or irrigation should be planned as an integral part of an overall farm system to conserve soil and water resources. Boundaries, elevations and direction of slope of individual field grading jobs should be such that the requirements of all adjacent areas in the farm unit can be met. Designs for the area being graded should include plans for removing excess irrigation and storm runoff water from the fields.

Excavation and fill material required for or obtained from such structures as ditches, pads and roadways should be planned for as a part of the overall grading job and the yardage included when calculating cut-to-fill ratios. The cut-to-fill ratio will normally be between 1.30 and 1.50 to allow for losses due to compaction, hauling and undercutting.

Furrow Grades

Land graded for irrigation with subsequent drainage will have a slope in the row direction between 0.1 percent and 0.5 percent on deep alluvial soils. There should be no reverse grade in the row direction. The most desirable surface is a plane. Fields graded to minimum slopes will require more maintenance of grade than steeper slopes.

Design grades on prairie claypan soils may have furrow grades up to 1.0 percent to avoid exposing large areas of subsoil in cut areas. Special residue management may be necessary to minimize erosion where furrow grades exceed 0.5 percent.

Cross Slope

Cross slope (slope perpendicular to row slope) is permitted in order to reduce cut yardage or to establish the "plane of best fit." Cross slopes must be such that "breakthroughs" from both irrigation water and runoff from rainfall are held to a minimum. Recommended cross slope is shown in table.

Table: Maximum recommended cross slope.

Furrow grade	Cross slope
0.1 percent	0.3 percent
0.2 percent	0.3 percent
0.3 percent	0.3 percent
0.4 percent	0.4 percent
0.5 percent	0.5 percent

On prairie claypan soils. cross slopes up to 3 percent are permitted. Use terraces on cross slope of 2 percent or more. Where terraces are necessary on fields to be irrigated, rows should be parallel to terraces. Land forming may be necessary between terraces to eliminate reverse grade in irrigated rows.

Maximum Length of Runs for Irrigation

Maximum length of runs for irrigation should be limited by furrow flow rates available, furrow cross-sectioned area, erosion hazard to the furrow, and water intake characteristics of the soil. Erosion hazard is a function of soil texture, crop residue and slope. A frequently used guide to maximum furrow flow rates is $Q = 10/s$, where s is furrow slope in percent and Q is gallons per minute per furrow.

An upper limit of 50 gpm per furrow is usually set due to furrow cross-section limitation.

Reasonable increments of length of run are fractional subdivisions of a mile due to land ownership patterns. The maximum length of run for irrigation is generally 1/4 mile.

The following table gives recommended maximum length of runs for various row slopes and soil textures. A range of run lengths is provided for some conditions. Use the shorter length of run for 2-inch irrigation applications. A longer run will result in poor water distribution through seepage and runoff. Maximum row lengths of 1/8 mile (660 feet) to 1/6 mile (880 feet), should be used on soils with high permeability or high erodibility. Soils such as Beulah Loamy Sand or Sarpy Loamy Sand have intake rates too high to be furrow irrigated. Drainage field ditches, sometimes referred to as tail ditches, should be properly designed to carry the excess flow from the furrows within reasonable time so as not to damage field crops.

Construction Specifications

Land to be graded should be cleared of brush and excessive crop residue, trash or vegetative material. Grading should not be attempted when soil moisture exceeds that permitting normal tillage or plowing.

Bring the land to design grade or grades in accordance with a detailed plan showing cuts, fills and grades. Fills of more than 6 inches should be built up by spreading the soil in successive layers. Disk or chisel the field surface after scoop work is completed and before final land planing.

Finish work with a land plane will be done so the field is free from depressions that would cause ponding of water. The land plane should be operated over the field three times: once at a 45 degree angle to the direction of the rows; once at a right angle to the direction of the rows; and finally in the direction of the rows. Field checking to determine compliance with design grades should have a maximum tolerance of plus or minus 0.1 foot at any grid point, with no reverse grade.

During the first year after grading, normally cut areas swell and fill areas settle. This may require minor cuts-and-fills and additional land planing.

Table: Recommended maximum length of run in feet for 2- to 3-inch application.

Row grade	Maximum furrow stream GPM (Q)	Length of run (feet)[1]			
		Soil texture			
		H[2]	F[3]	M[4]	S[5]
0.1 percent	50	1320	1320	1320	800 to 1320
0.25 percent	40	1320	800-1320	880	660 to 880
0.75 percent	13	660-880	660	660	
1 percent	10	660	660		

[1]The run lengths shown in the table also are applicable to border irrigation. Length of run and slope may be increased for erosion-resistant grass or grass-legume crops. Maximum cross slope should not exceed 0.1 foot per border strip width.

[2]H — Fine textured (sandy clays, silty clays and clays): Typical soils in this group are Sharkey clay, Osage clay. Carlow silty clay, and Wabash silty clay.

[3]F — Moderately fine textured (sandy clay loams, clay loams and silty clay loams): Typical soils in this group are Onawa silty clay loam, Zook silty clay loam and Colo Silty clay loam.

[4]M — Medium textured (very fine sandy loams, loams and silt loams): Typical soils in this group are Putman silt loam, Parsons silt loam, Robinsonville fine sandy loam, Dundee loam, Westerville silt loam, Nodaway silt loam and Blackoak silt loam.

[5]S — Moderately coarse textured (Sandy loams and fine sandy loams): Typical soils in this group are Bruno sandy loam and Bosket fine sandy loam.

Land Levelling

Levelling, smoothing and shaping the field surface is as important to the surface system as the design of laterals, manifolds, risers and outlets is for sprinkler or trickle irrigation systems. It is a process for ensuring that the depths and discharge variations over the field are relatively uniform and, as a result, that water distributions in the root zone are also uniform. These field operations are required nearly every cropping season, particularly where substantial cultivation following harvest disrupts the field surface. The preparation of the field surface for conveyance and distribution of irrigation water is as important to efficient surface irrigation as any other single management practice the farmer employs.

There are perhaps two land levelling philosophies:

1. To provide a slope which fits a water supply; and

2. To level the field to its best condition with minimal earth movement and then vary the water supply for the field condition. The second philosophy is generally the most feasible.

Because land levelling is expensive and large earth movements may leave significant areas of the field without fertile topsoil, this second philosophy is also generally the most economic approach.

Land levelling always improves the efficiency of water, labour and energy resources utilization. The levelling operation, however, can be the most intensively disruptive cultural practice applied to the field and several factors should be considered before implementing a land levelling project. Major topographical changes will nearly always reduce crop production in the cut areas until fertility can be replaced. Similarly, equipment traffic can so compact or pulverize the soil that water penetration is a major problem for some time. The farmer has many activities which contribute to his productivity and therefore require his skill and labour. The irrigation system should be designed with him (or her) in mind. A field levelled to high standards is generally more easily irrigated than one where undulations require special attention.

New equipment is continually being introduced which provides the capability for more precise land levelling operations. One of the most significant advances has been the adaptation of laser control in land levelling equipment. The equipment has made level basin irrigation particularly attractive since the final field grade can be very precise. Comparisons with less precise techniques have clearly shown that laser-levelled fields achieve better irrigation and production performance. Nevertheless, for most irrigated agriculture, laser-controlled precision is unfeasible because of the high cost of such equipment unless a large number of farmers form a cooperative or a government programme is started with subsidized land levelling as one component in an effort to improve farm production.

Small-scale Land Levelling

Most small-scale farming operations rely on animal power or small mechanized equipment which an individual can own and operate. As the irrigator waters his fields season after season he is able to observe the locations of high and low spots on the field. Then as he prepares the fields between plantings, he tries to move soil from the high spots to the low ones. Over a period of several years individual fields are smoothed enough to be watered fairly well. Figures show two examples of these operations. In figure a farmer is preparing land for paddy and using the ponded water level on the field to direct him to the high and low spots. Since this is a normal land preparation practice, it does not represent an extra task for the irrigator. Figure shows a similar operation using mechanized equipment for typical annual crops and again one sees that the field preparation also readies the seed bed for planting. Beyond these technologies one may observe various levels of mechanization and an array of implements. The one feature common to most small-scale land levelling is the trial and error nature of the practices and the long-term incorporation of land levelling with seed bed preparation. Another feature is that no technical or engineering inputs are needed.

Traditional Engineering Approach

Initial Considerations

Initially, the field should be studied and an overall irrigation strategy identified. Once accomplished, the land levelling programme derived from traditional engineering practice can be initiated. The first step is to establish the plane of the field. This involves placing a reference grid on the field, surveying the existing topography of the field by establishing the elevations of the grid points, and calculating the new field topography by adjusting the grid elevations to correspond to the desirable plane. This is the engineering phase of the land levelling procedure. Once the surface design has been determined, a land levelling operation begins. This is typically a private contractor utilizing his equipment to move the earth into the new position on the field, and the adequacy of the land levelling is dependent on the skill of the equipment operator.

Engineering Phase

Surveying and mapping the field involves setting a uniform grid system on the field and establishing the field topography. This need not be a complicated procedure. One corner of the field can be chosen as a starting point and the first stake can be located one-half grid spacing from either boundary. Then a row of stakes can be measured and set using a transit or level and tape. The instrument is set up over the first stake and sighted along a line parallel to the boundary. Usually this is accomplished by going to the opposite edge and locating a stake one-half grid spacing from the edge. Then, using the instrument for alignment, the first row of stakes is measured into place. With the instrument located over the same stake and aligned along the first row, the next step is to turn the alignment 90° by either measuring a right triangle or by using the instrument angle indicators if available. The new alignment is used to locate another stake row along the other field axis. Each of the remaining stakes can be placed visually by sighting against the two stakes at the field edges. The grid spacing can be set at convenient lengths so long as it is square and consistent (this is not technically required but it simplifies the calculations). In the US, the typical grid spacing is 100 feet by 100 feet (30.5 m by 30.5 m). However this would be too large in many countries with small fields. It is suggested that the surveyor use a multiple of 10 m as a spacing and select one that divides the field into at least 5 percent subareas.

The field stakes provide the basis of the field survey. The level or transit can be located in a central area and rod readings taken from each stake position. It is generally advisable to locate a benchmark near the field from which to reference the readings as elevations. In addition, readings taken from the location of water supply structures are also useful for designing the head ditches, watercourses and drainage channels. It is assumed that the basic principles of land surveying are known and practiced during this phase of the land levelling operation.

An initial decision as to the method of surface irrigation will dictate field slope. Basins are designed to be level in both field directions. Borders are similar in having zero cross-slope, but may have advance slopes of up to 2 or 3 percent, depending on crop and soil conditions. Furrow irrigation systems work well with advance slopes up to 1 to 3 percent and cross-slopes of 0.5 to 1.5 percent. If the average natural slopes are greater than these ranges, terraces or benches should be planned.

There are several ways to compute the new field slope including some that are inspection methods requiring some experienced judgment. A formal method, called the 'plane method,' will be used here.

The plane method is a simple least squares or linear regression fit of field elevations to a two-dimensional plane. Subsequent adjustments are made in the elevation of the plane centroid to compensate for variable cut/fill ratios. If the field has a basic X-Y orientation, the plane equation is written as:

$$E(X, Y) = AX + BY + C$$

in which:

E = elevation of the X, Y coordinate;

A, B = regression coefficients; and

C = elevation of the origin or reference point for the calculations of field topography using equation.

The first step in evaluating the constants, A, B and C, is to determine the weighted average elevations of each grid point in the field. The purpose of the weighing is to adjust for any boundary stakes that represent larger or smaller areas than given by the standard grid dimension. The weighing factor is defined as the ratio of actual area represented by a grid point to the standard area. The grid point area is assumed to be the proportional area surrounding the stake or other identification of the grid point elevation. The weighing factor is:

$$\theta_{ij} = \frac{A_{ij}}{A_s}$$

where:

θ_{ij} = weighing factor of the grid point identified as the ith stake row and the jth stake column;

A_{ij} = area represented by the (i, j) grid point; and

A_s = area represented by the standard grid dimension.

The next step is to determine the average elevation of each row and column. For the ith row, E_i, is:

$$E_i = \frac{\sum\limits_{j-1}^{N'} \theta_{ij} E_{ij}}{\sum\limits_{j-1}^{N'} \theta_{ij}}$$

in which:

N' = number of stake columns; and

E_{ij} = elevation of the (i, j) coordinate found from field measurements E(X, Y).

A similar expression can be written for finding the average elevation of the jth stake column, E_j:

$$E_j = \frac{\sum\limits_{i-1}^{N''} \theta_{ij} E_{ij}}{\sum\limits_{i-1}^{N''} \theta_{ij}}$$

where N" is the number of stake rows.

The next step is to locate the centroid of the field with respect to the grid system. For convenience, an origin can be located one grid spacing in each direction from the first stake position, i.e. the initial stake position on the field. The distance from the origin to the centroid in the X dimension is found by:

$$X = \frac{\sum\limits_{j-1}^{N'} \theta_j X_j}{\sum\limits_{j-1}^{N'} \theta_j}$$

where:

X = x distance from origin to centroid;

X_j = x distance from origin to the jth stake column position; and

$$\theta_j = \sum\limits_{i-1}^{N''} \theta_{ij}$$

Similarly,

$$Y = \frac{\sum\limits_{i-1}^{N''} \theta_i Y_i}{\sum\limits_{i-1}^{N''} \theta_i}$$

in which:

Y = y distance from the origin to centroid;

Y_i = y distance from origin to the ith stake row position; and

$$\theta_i = \sum\limits_{j-1}^{N'} \theta_{ij}$$

The fourth step is to compute a least squares line through the average row elevations in both field directions. The slope of the best fit line through the average X-direction elevation (E_j) is A and is found by:

$$A = \frac{\sum\limits_{j=1}^{N'} X_j E_j - \left(\sum\limits_{j=1}^{N'} X_j\right)\left(\sum\limits_{j=1}^{N'} E_j\right)/N'}{\sum\limits_{j=1}^{N'} X_j^2 - \left(\sum\limits_{j=1}^{N'} X_j\right)^2/N'}$$

For the best fit slope in the Y-direction, the slope, B, is:

$$B = \frac{\sum\limits_{j=1}^{N''} X_i E_i - \left(\sum\limits_{j=1}^{N''} X_i\right)\left(\sum\limits_{j=1}^{N''} E_i\right)/N''}{\sum\limits_{j=1}^{N''} Y_i^2 - \left(\sum\limits_{j=1}^{N''} Y_i\right)^2/N''}$$

Finally, the average field elevation, E_F, can be found by summing either E_i or E_j and dividing by the appropriate number of grid rows. This elevation corresponds to the elevation of the field centroid (X, Y). Thus, equation $E(X, Y) = AX + BY + C$ can be solved for C as follows:

$$C = E_F - {}_A X - {}_B X$$

An adjusted elevation for each stake can be computed with Equation $S_f = y_o/x$ and compared to the measured values. The differences are the necessary cuts or fills. Before these computations are undertaken, however, the slopes in both field directions must be checked to see if they are within satisfactory limits. For example, if the intended system is a border irrigation system, the cross-slope should be zero $(A = 0)$ and the cuts and fills would need to be based on this condition. A second note concerns the fact that cuts and fills do not balance because of variations in soil density.

Adjusting for the Cut/Fill Ratio

In most cases, the best fit plane and the subsequent adjusted elevation will result in different total volumes of cuts or fills. A simple and rapid calculation of these respective volumes can be made as follows:

$$V_c = \sum_{m-1}^{Nc} A_m C_m$$

and,

$$V_f = \sum_{n-1}^{Nf} A_n F_n$$

in which:

V_c = volume of cuts, m³;

V_f = volume of fills, m³;

A = grid area m or n, m³;

C_m = depth of cut at grid point m, in metres, and

F_n = depth of fill at grid point n, in metres.

The cut/fill ratio r is:

$$r = V_c / V_f$$

and should be in the range of 1.1 to 1.5 depending on the soil type and its condition.

The necessity of having cut/fill ratios greater than one for land levelling operations stems from the fact that in disturbing the soil, the density is changed (the fill soil is more dense because its structure has been destroyed). Selecting a cut/fill ratio remains a matter of judgement. If the value arrived at by least squares is not in the range of 1.1 to 1.5, the elevation of the field centroid, C, is raised or lowered until the value of r is appropriate. This adjustment is determined by:

$$d = \frac{r'^* V_f - V_c}{\left(\sum_{i-1}^{Nc} A_i\right)(1+r')}$$

where r' is the cut/fill ratio required in the design.

Equations r = V_c / V_f and above equation assume that none of the 'cut' grid points become 'fill' points or vice-versa. Consequently, in some cases it will be necessary to iterate a few times to get the proper cut/fill ratio.

Equation $V_c = \sum_{m-1}^{Nc} A_m C_m$ is usually less formal than required for contracting purposes. Some more complete estimators include the prismoidal formula, the 'average end area method,' and the 'four corners method.' The 'four corners method' is simplest to use and is suggested by the USDA (1970). The formula for all complete grid spacings is:

$$V_c = \frac{A_i}{4} \cdot \frac{\left(\sum_{j-1}^{Nc} C_j\right)^2}{\sum_{j-1}^{Nc} C_j + \sum_{m-1}^{4.Nc} F_m}$$

in which:

   A_i = area of the grid square i, m²;

   N_c = number of cuts at the four corners of the grid square; and

   C_j and F_m = cut and fill depths in m, but they are taken as absolute values so they both have the same sign, positive.

At the field edges and corners, if complete grid spacings are not present, the cut volume must be computed separately. The procedure is to assume the elevations of the field boundaries are the same as the nearest stake and would thereby have the same cut or fill dimensions. Above equation is then utilized with appropriate A_i value corresponding to the actual edge area.

Some Practical Problems

The engineering design derived from the procedures above results in a field design which should provide the irrigator with a system that will satisfy his irrigation practices and yield efficient and uniform waterings if managed properly. Between the design and the operable system is the land levelling operation itself. Generally, a contractor must be retained to move the earth, after which the field topography is checked and if necessary the contractor refines his job with additional work. The skill and efficiency of the equipment operator is critical to how well the field levelling is finally accomplished. A good operator may be able to provide a field grade within plus or minus 10 cm; a poor operator perhaps double this value. The first of the practical problems is the arrangements between the irrigator and the contractor. The work should be checked and fall within the 10 cm limits before it is accepted and reimbursed.

Land levelling is likely to be not only the most disruptive operation applied to the field but also the most costly. One method of reducing cut volumes, and therefore the cost, is to subdivide the field into terraces or benches. Usually, earthwork is minimized when the terrace runs parallel to the direction of highest field slope but to be sure, the cut volumes should be checked with the alternative field layouts.

Operators develop field movement patterns based on their own judgement and experience. A cut-haul-fill pattern of travel that maximizes the efficiency of the land levelling operation tends to be one in which the routes are of nearly equal length. Such a strategy prevents the over-use of travel lanes and minimizes the haul and return distances. Where manually controlled equipment is used, many operators establish a bench mark grid over the field by cutting and filling strips on both sides of a stake to the desired grade. Then the median areas can be levelled to grade to better precision. Good operators make cut and fill passes which are relatively uniform and their equipment is seen to operate at fairly uniform speeds, particularly during loading passes.

Earth may be used to raise the elevation of roadways, or prepare a raised pad for headland facilities. In the computation setting field cuts and fills, the volume of the earth needed for these miscellaneous requirements should be deducted in the cut/fill ratio calculation.

The topography of surface irrigated fields, even after levelling, is not a static feature of the land. Year to year variations in tillage operations disturb the surface layers as well as shift their lateral position. The loose soils may settle differently depending upon equipment travel or depths of irrigation water applied. Consequently, a major land levelling operation will correct the macro-topographical problems but annual levelling or planing is needed to maintain the field surface by correcting micro-topographical variations.

Booher devotes a chapter in his manual on surface irrigation to land levelling. Included is an example problem around which useful suggestions are made regarding the methods and equipment for levelling the field into a workable surface irrigated field. The problem that is developed utilizes a different approach to that suggested herein so it will be partially repeated for purposes of both illustration and comparison.

The first six columns and the first eight rows of Booher's example field have been extracted and are shown in figures. The locations of the field boundaries have been changed relative to the grid system to illustrate the importance of weighing grid point elevations based on the areas they represent. In the following example the standard grid spacing is 20 m by 20 m and begins one-half spacing from the upper left corner of the field (represented by the grid point [i, A] in figure). The standard grid area is 400 m_2, but one will note that grid points adjacent to the right field boundary represent 500 m². One point, the lower right grid represents an area of 375 m².

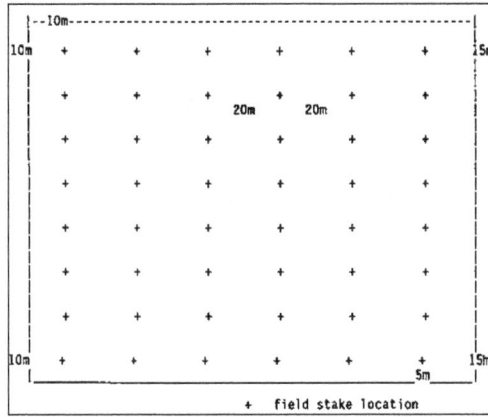

Example problem field layout.

Initial field elevations in metres.

The first step in the calculation of the revised field plane is to determine the grid point weighing factors using Eq. $\theta_{ij} = \dfrac{A_{ij}}{A_s}$. Using the standard area per point as 400 m², the weighing coefficients, θ_{ij}, are shown in figure. The row and column weights are the sum of the grid point weights and are shown to the left and at the bottom of figure.

Grid point weighing coefficients.

Using the column and row weights, Eqs. $E_i = \dfrac{\sum\limits_{j\text{-}1}^{N'} \theta_{ij} E_{ij}}{\sum\limits_{j\text{-}1}^{N'} \theta_{ij}}$ and $E_j = \dfrac{\sum\limits_{i\text{-}1}^{N''} \theta_{ij} E_{ij}}{\sum\limits_{i\text{-}1}^{N''} \theta_{ij}}$ are used to calculate the

average elevation of the respective rows and columns. These data are included along the left and bottom of figure.

The field centroid is calculated with Eqs. $X = \dfrac{\sum\limits_{j\text{-}1}^{N'} \theta_j X_j}{\sum\limits_{j\text{-}1}^{N'} \theta_j}$, $Y = \dfrac{\sum\limits_{i\text{-}1}^{N''} \theta_i Y_i}{\sum\limits_{i\text{-}1}^{N''} \theta_i}$ and $\theta_i = \sum\limits_{j\text{-}1}^{N'} \theta_{ij}$ using the dis-

tances from the origin and the row and column weights. For the X coordinate of the centroid, this calculation is:

$$X = \frac{7.75\left(20+40+60+80+100\right)+120^*\left(9.688\right)}{5^*7.75+9.688} = 72\,\text{m}$$

and for the Y coordinate:

$$y = \frac{6.25^*\left(20+40+60+80+100+120+140\right)+4.688^*\left(160\right)}{7^*6.25+4.688} = 87.743\,\text{m}$$

Note that the origin is 10 m to the right and 10 m above the stake at grid position [i, A].

The next step is to run a linear regression through the average row and column elevations using

Equations $A = \dfrac{\sum\limits_{j\text{-}1}^{N'} X_j E_j - \left(\sum\limits_{j\text{-}1}^{N'} X_j\right)\left(\sum\limits_{j\text{-}1}^{N'} E_j\right)/N}{\sum\limits_{j\text{-}1}^{N'} X_j^2 - \left(\sum\limits_{j\text{-}1}^{N'} X_j\right)^2/N}$ and $B = \dfrac{\sum\limits_{j\text{-}1}^{N''} X_i E_i - \left(\sum\limits_{j\text{-}1}^{N''} X_i\right)\left(\sum\limits_{j\text{-}1}^{N''} E_i\right)/N}{\sum\limits_{j\text{-}1}^{N''} Y_i^2 - \left(\sum\limits_{j\text{-}1}^{N''} Y_i\right)^2/N}$.

These procedures are fairly standard on hand-held calculators and microcomputers so the calculations will not be shown here. The slope of the field from right to left is 0.000373 (A) and that from top to bottom is -0.002247 (B). It can also be mentioned that standard regression techniques will also yield an intercept value representing the elevation with which the best fit line through the average elevations will intercept the X and Y axis running through the origin. These intercepts can be ignored.

The final calculations involving the revised field plane involve the calculation of the C value in Eq. 117 as outlined in the paragraph preceding Eq. C = EF - A X - B X. The average elevation at the centroid of the field is determined by summing the average row or column elevations. This value is also shown in figure as 1.557 m. From Eq. C = EF - A X - B X, then:

C = 1.577 - 0.000373 * 72 - (-.002746 * 87.743) = 1.7911 m

The resulting equation of the field plane defined by the procedure so far is:

E(X, Y) =.000373 * X -.002746 * Y + 1.7911

If this relationship is used to recompute the elevations at each grid point, the cuts and fills are identified as the positive (fills) or negative (cuts) differences between the computed elevations and the original topography. Figure shows these results as the upper number near the grid points.

First determination of cuts and fills for the example problem.

In order for the earthwork to balance in the field after levelling, the volume of cuts should exceed the fills by 10 to 30 percent. For the 6th row shown below, Eqs. $V_c \sum_{m-1}^{Nc} A_m C_m$ and $V_f = \sum_{n-1}^{Nf} A_n F_n$ are evaluated as follows:

	+.11	+.03	-.01	-.01	-.02	0
vi	*	*	*	*	*	*

Volume of Cuts for Row vi = (-.01) * 400 + (-.01) * 400 + (-.02) * 400 = -16 m³

or since the sign is irrelevant, the cut volume along row 6 is 16 m³, and for the fills:

Volume of Fills for Row vi = 400 * (.11 +.03 +.000) = 56 m³

Determining the cuts and fills of each row and then summing yields a total cut volume of 627 m³ and a total fill volume of 1007 m³. Dividing the cut volume by the fill volume gives a cut/fill ratio of about 0.62, which of course is not satisfactory.

Assuming the cut/fill ratio should be about 1.3, Equation $d = \dfrac{r'^* V_f - V_c}{\left(\sum_{i-1}^{Nc} A_i\right)(1+r')}$ can be used to re-compute the elevation of the field centroid, **C**. The change in centroid elevation is determined by summing the area of each cut station times the depth of cut. There are 17 cut points in which the grid area is 400 m², 2 involving the 500 m² left boundary points, and 4 cuts along the 300 m grid points along the lower field boundary. Thus the area summation in the denominator of Equation

$\theta_i = \sum_{j-1}^{N'} \theta_{ij}$ is 9000 m². The remainder of Equation $\theta_i \sum_{j-1}^{'} \theta_{ij}$ is then:

$$d = \frac{1.3^*1007 - 627}{9000^*\left(1+1.3\right)} = 0.033\,\text{m}$$

This calculation assumes that none of the previous fill locations become cut locations. To test this assumption, 0.033 m is subtracted from each cut and fill depth in figure and the results are shown in figure. It is noted that 2 fill locations have become cut points.

Second determination of cuts and fills for the problem.

Recomputing the volume of cuts from Eq. $V_c = \sum_{m-1}^{Nc} A_m C_m$ and the fills from Eq. $V_f = \sum_{n-1}^{Nf} A_n F_n$ yields the following cut/fill ratio (Eq. $r = Vc / Vf$):

$$\text{cut/ fill ratio} = \frac{\text{cut volume}}{\text{fill volume}} - \frac{959\,m3}{708\,m3} = 1.35$$

This value is slightly more than the 1.3 assumed in adjusting the C value in Eq. $E(X, Y) = AX + BY + C$ and reflects the problem of grid points changing from cuts to fills (or vice versa in other cases). If the error had been greater, another iteration would be suggested. Not in this case, however, and the final field plane is as shown in figure with the subscript cuts and fills.

If the field is intended for borders and basins, the procedure is the same except that the A and/ or B slopes in Eq. $E(X, Y) = AX + BY + C$ would be zero. Similarly, if the field is to be terraced, the procedure is applied separately to the grid points in each terrace area.

The last engineering step is the formal computation of the volume of cuts for contractual purposes.

This is illustrated for the evaluation of Eq. $V_c = \frac{A_i}{4} \cdot \frac{\left(\sum_{j-1}^{Nc} C_j\right)^2}{\sum_{j-1}^{Nc} C_j + \sum_{m-1}^{4.Nc} F_m}$ for the area between rows v and

vi. The final cut/fill depths for rows vii and viii are shown below:

v		*	*	*	*	*	*	
		+.28	+.18	+.05	+.01	0	+.05	
vi		*	*	*	*	*	*	
		+.08	-.01	-.04	-.04	-.05	-.04	

It is assumed that the depth of fill at the left boundary is .28 m at row v and .08 m at row vi. Similarly, the fill and cut at the right boundary are .05 m at row v and -.04 at row vi respectively. Equation

$$V_c = \frac{A_i}{4} \cdot \frac{\left(\sum_{j-1}^{Nc} C_j\right)^2}{\sum_{j-1}^{Nc} C_j + \sum_{m-1}^{4.Nc} F_m}$$ is evaluated as follows:

Grid Points		Computations	Total	
|	*			
+.28	+.28	$\dfrac{200\,m^2}{4} * \dfrac{(0)2}{.28+.28+.08+.08}$	=	0 m³
|	*			
+.08	+.08			
*	*			
+.28	+.18	$\dfrac{400\,m^2}{4} * \dfrac{(-.01)2}{abs(-.01)+.28+.08+.18}$	=	.02 m³
*	*			
+.08	-.01			
*	*			
+.018	+.05	$\dfrac{400}{4} * \dfrac{(-.01+-.04)2}{abs(-.01+-.04)+.28+.18+.05}$	=	.89 m³
*	*			
-.01	-.04			
*	*			
+.05	+.01	$\dfrac{400}{4} * \dfrac{(-.04+-.04)2}{abs(-.04+-.04)+.05+.01}$	=	4.57 m³
*	*			
-.04	-.04			
*	*			
+.01	+.0	$\dfrac{400}{4} * \dfrac{(-.04+.05)2}{üüüüüüü(-\quad +- \quad)+ \quad +}$	=	8.10 m³
*	*			
-.04	-.05			
*	*			
0	+.05	$\dfrac{400}{4} * \dfrac{(-.04+-.05)2}{abs(-.04+-.05)+.05}$	=	5.79 m³
*	*			
-.05	-.04			

*						
+.05	+.05					
		$$\frac{400}{4} * \frac{(2^* - .04)2}{\text{abs}(2^* - .04) + 2^*.05}$$		=	4.44 m³	
*						
-.04	-.04					
				Total	23.81 m³	

Repeating these calculations for each grid area yields a total cut volume of 946.02 m³ which is very close to the 959 m³ estimated with Eq. $V_c = \sum_{m-1}^{N_c} A_m C_m$.

It is perhaps worthwhile mentioning at this point that microcomputer programmes have been written to perform land levelling computations as illustrated above. Some of these are commercially available, some can be acquired by tracking down the programmer.

Laser Land Levelling

The advent of the laser-controlled land levelling equipment has marked one of the most significant advances in surface irrigation technology. One such system is shown in figure. It has four essential elements:

1. The laser emitter;
2. The laser sensor;
3. The electronic and hydraulic control system; and
4. The tractor and grading implement.

Land levelling equipment

The laser emission device involves a battery operated laser beam generator which rotates at relatively high speed on an axis normal to the field plane. This rotating beam thereby effectively creates a plane of laser light above the field which can be used as the levelling reference rather than the elevation survey at discrete grid points in conventional land levelling techniques. Various beam generators are equipped with self-adjustment mechanisms that allow the plane of the beam to be aligned in any longitudinal or latitudinal slope desired. This reference plane of laser light is an extremely advantageous factor in the levelling operation because it is not affected by the earth movement, does not require a field survey to establish the high and low spots, and does not require the operator to judge the magnitude of cuts and fills. The distance between the laser beam and the earth surface is defined such that deviations from this distance become the cuts and fills. With laser systems, there is little or no need for the exhaustive engineering calculations of the conventional approach. The cost of levelling is usually contracted on the basis of money per equipment hour. The laser emitter is generally located on a tripod or other tower-like structure on or near the field and at an elevation such that the laser beam rotates above any obstructions on the field as well as the levelling equipment itself. The beam is targeted and received by a light sensor mounted on a mast attached to the land grading implement. The sensor is actually a series of detectors situated vertically so that as the grading implement moves up or down, the light is detected above or below the centre detector. This information is transmitted to the control system which actuates the hydraulic system to raise or lower the implement until the light again strikes the centre detector. It is in this manner that the sensor on the mast is continually aligned with the plane on the laser beam and thereby references the moving equipment with the beam. It is important to note that the sensitivity of the laser sensor system is at least 10 to 50 times more precise than the visual judgement and manual hydraulic control of an operator on the tractor. Consequently, the land levelling operation is correspondingly more accurate. The skill of the operator is substantially less critical to the levelling which allows farmers and other personnel access to the land grading equipment.

Close up View of Laser Beam Emmitter

The electronic and hydraulic control systems generally have two operating modes. In the first, or observation mode, the mast itself moves up or down according to the undulations in the field as the operator drives the equipment over the field in a grid-like fashion. The monitor in the tractor yields elevation data from which the operator can determine average field elevations and slopes. In other words, the system operates as a self-contained surveying system. In this mode, the blade of the grading implement is fixed in place and only the sensor mast moves. In the second mode, or planing mode, the mast position is fixed relative to the implement blade which is then raised or lowered in response to the land topography. The beam plane is located the appropriate distance above the field centroid and at the desired slopes. By adjusting the height of the mast sensor relative to this plane and the centroid, the cutting and filling is accomplished simply by driving the tractor over the field. However, in many cases, the depth of cuts will exceed the depth which can be cut with the power of the tractor and the operator must override the automatic controls in order to keep the equipment operating.

The fourth element of the levelling system is the tractor - grading implement combination. This equipment is generally standard agricultural tractors and land graders in which the hydraulic and control systems have been modified to operate under the supervision of the electronic controller supplied with the laser emitter and sensor devices. The tractor needs to be carefully selected so that it is not under-powered and its hydraulic system is strong enough to work with the laser-imposed frequency of movements and adjustments. The grading implement can be as simple as a land plane which scrapes the earth and moves only as much as can be pushed in front of the blade or a complex piece of equipment which loads and carries earth. The former is used primarily for small levelling jobs, smoothing and repeat grading. The latter is usually better for initial levelling where cuts are larger and in the preparation of level basins where the cuts are also larger than in bordered or furrowed fields.

As a final note on levelling in general and laser levelling is particular, it is probable that the importance of accurate field grading has been under estimated. The precision improves irrigation uniformity and efficiency and as a result the productivity of water and land. On large fields, the improved productivity has been shown to pay economic dividends that easily exceed the cost of the levelling. However, the equipment is expensive and quite beyond all but the largest of farmers. In the developing countries, laser-guided equipment is being demonstrated and tested. There remains the solution as to how such equipment can be made useful for the small farmer.

Contour Bench Levelling and Earthwork Quantities Computation

Irrigation in undulating fields and steep slope is a very difficult task. Rainfall erosion and moisture conservation can be controlled and retained by forming contour benches. Earthwork computation is an important task in land levelling.

Contour Bench Levelling

Contour bench levelling is a method for cutting length of slope to a desired grade and preparing land for irrigation. The undulated field is cut into a number of steps approximately by using contours; each step is levelled and made as an independent area. Thus a series of steps are formed in successive elevations around the slope. Benches are used for forming the border, furrow and check basin the slope. The contour bench levelling provides controlled irrigation water flow on the flat slopes and for efficient irrigation. Contour bench levelling controls erosion from rainfall, and permits soil building processes thereby resulting in increase of fertility and improved soil structure. The flat benches provide greater opportunity time for infiltration thereby reducing the quantity of irrigation water needed to meet plant requirements.

Construction of Contour Benches

The components of bench cross-section are shown through figure. Selecting the proper cross section for contour benches is one of the most important steps of planning for contour bench construction.

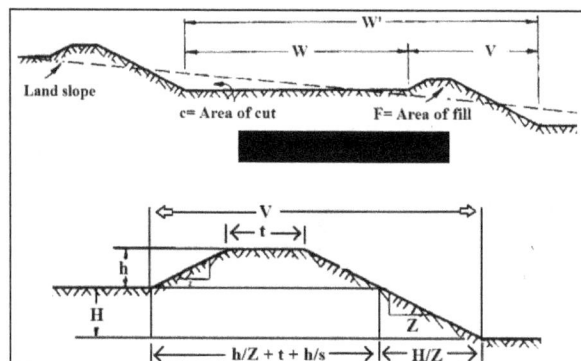

Cross section of bench and bund.

Let W be the bench width of the farmable area. This width should be such that it can accommodate the widest of farm equipment to be used for farming.

The other parameters are:

W' = overall bench width (m)

W = width of cultivable strip (m)

t = top width of dike (m)

h = height of dike (m)

H = vertical interval between benches (m)

Z = side slope of dike (dimensionless)

S = slope of land (dimensionless)

The relationship between W, W$^-$ and H can be expressed as:

W$^-$ = H/S

or, H = W$^-$ S

$$W' = W + \frac{h}{Z} + t + \frac{h}{Z} + \frac{W'S}{Z}$$

$$W' - \frac{W'S}{Z} = W + t + \frac{2h}{Z}$$

$$W'\left(1 - \frac{s}{z}\right) = W + t + \frac{2h}{Z}$$

$$W' = \left(W + t + \frac{2h}{Z}\right)\left(\frac{1}{1 - s/z}\right) = \left(W + B + \frac{2h}{Z}\right)\left(\frac{z}{z - s}\right)$$

The side slopes of dike (Z) should have stable side slopes. The side slope of 2:1 is normally provided. The area where stones are presents in the field much steeper slope can be used. Stones should be used to support the bund. Top width of the dike, (t) should be sufficient to prevent further lowering of its height by trampling or by other sources.

It is a usual practice to keep the top width of dike equal to the vertical interval between benches:

t = H

Example: A trapezoidal bund of 80 m long is to be constructed having bottom width as 4 m and top width as 2 m. The height of one end of bund is 1.2 m and that of the other end is 1.5 m. Determine the volume of earth fill for making bund.

Solution:

The volume of earth fills for making bund $(V) = \frac{(A_1 + A_2)}{2} \times L$

Area of one end of bund (A_1) = ½(4+2) 1.2 =3.6 m²

Area of other end of bund (A_2) =½(4+2) 1.5 = 4.5 m²

$$V = \frac{(3.6+4.5)}{2} \times 80$$

V = 324.00 m³

Hence volume of earth fill = 324.00 m³

Example: The random field ditch drains are to be used for removal of drainage water. The plan, profile and cross section are shown Fig. Estimate the volume of earth work for cutting.

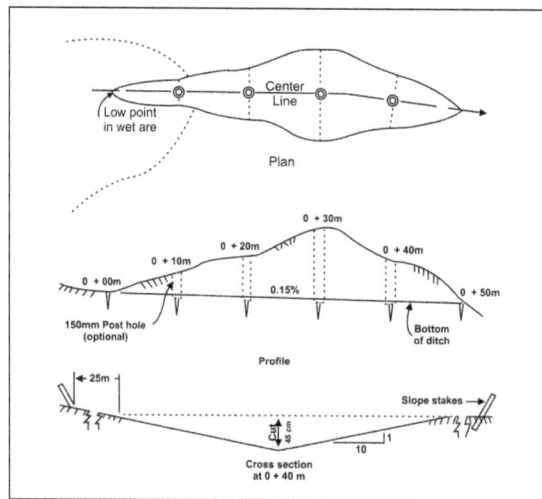

Layout of random field drain for computing the earth work in cutting.

Solution:

The levelling instrument was used to obtain the depth of cut for the ditch grade of 0.15 per cent. The procedure used to compute earth work is illustrated in the following table.

Station (m)	Cut (m)	Top Width (m)	Cross-sectional area (m²)	Average Cross-sectional area (m²)	Distance (m)	Volume of cut (m³)
0+00	0	0	0			
0+10	0.20	2.0	0.40	0. 2	10	2.00
0+20	0.35	3.5	1.23	0.82	10	8.20
0+30	0.70	7.0	4.9	3.07	10	30.70
0+40	0.45	4.5	2.03	3.47	10	34.70
0+50	0.00	0.0	0.0	1.02	10	10.20
Total						85.80

The total volume of earthwork in cutting is 85.8 m³.

Earthwork Quantities Computation

Earthwork quantities need to be computed for desired land levelling method and for generated cross section. The common methods for computing earthwork quantity are: end area method, prismoidal and four point method.

End Area Method

The areas of cuts and fills on the profiles or grid lines are used to compute the volume between the adjacent profile or grid lines, given by relationship:

$$V = L\frac{(A_1 + A_2)}{2}$$

In which,

V = volume of cut or fill, m³

L = distance between profiles or lines, m

A_1 = area of cut or fill in the first profile or line, m²

A_2 = area of cut or fill in the second profile, m².

Prismoidal Formula

A precise method of computing the volume of earthwork in land levelling makes use of the prismoidal formula:

$$V = \frac{L}{6}(A_1 + 4A_m + A_2)$$

In which,

V = volume of earthwork, m³

L = perpendicular distance between end planes, m

A_1 = area of the first end plane, m²

A_2 = area of the second end plane, m²

A_m = area of middle section parallel to end planes (m²)

Four – Point Method

A commonly used method called the four-point method is sufficiently accurate for land grading. Volume of cuts for each grid square is given by,

$$V_c = \frac{L^2 \left(\sum C\right)^2}{4\left(\sum C + \sum F\right)}$$

where,

V_c = volume of cut, m³

L = grid spacing, m

C =sum of cut on the four corners of a square grid, m

F = sum of fill on the four corners of a square grid, m

For computing V_f the volume of fills, $(\Sigma C)^2$ in the numerator of equation $V_c = \dfrac{L^2 (\Sigma C)^2}{4(\Sigma C + \Sigma F)}$ is replaced by (ΣF).

Example: Compute the balancing depth (volume of earthwork in cutting is equal to volume of earthwork in filling) for a canal having a bed width as 8 m with side slopes of 1:1 in cutting and 2:1 in filling. The bank embankments are kept 2 m higher than the ground level (berm level) and crest width of embankments is 2.0 m.

Solution:

The channel section is shown in figure. Let d_1 be the balancing depth, i.e. the depth for which excavation and filling becomes equal.

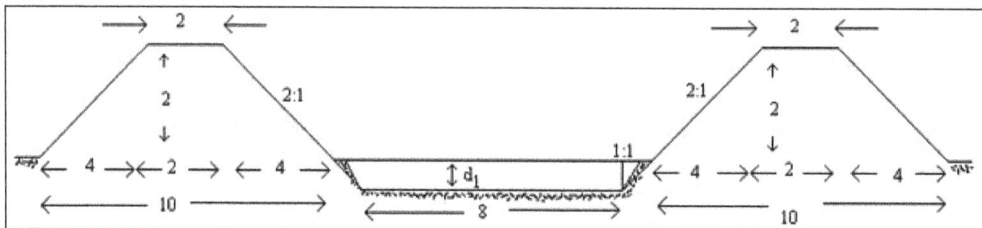

(All dimensions are in meter)
Cross section of a canal with embankment.

Let the length of canal be L meter.

Area of cutting $(A_1) = (8+d_1) d_1$

Volume of earthwork in cutting $V_1 = [(8+d_1).d_1].L$

Area of cross section of two embankments $= 2\left[\dfrac{2+10}{2} \times 2\right] = 24\,m^2$

Volume of earthwork in filling in construction of embankments = 24 L m²

Equating equations $(V_1 = [(8+d_1).d_1].L)$ and $\left(=2\left[\dfrac{2+10}{2} \times 2\right] = 24\,m^2\right)$, we get:

$[(8+ d_1) d_1]L = 24L$

or, $d_1 + 8 d_1 - 24 = 0$

$d_1 = \dfrac{-8 \pm \sqrt{64-4(-24)}}{2}$

$$= \frac{-8 \pm \sqrt{64 + 96}}{2}$$

$$= \frac{-8 \pm 12.65}{2}$$

$$d_1 = \frac{-8 + 12.65}{2} = 2.325\,\text{m}$$

$$d_1 = \frac{-8 - 12.65}{2} = -10.325\,\text{m}$$

Ignoring infeasible −ve sign, we get = 2.325 m.

Balancing depth = 2.325 m.

Chapter 5

Irrigation Canals

An irrigation canal is a man-made waterway to transport water for agricultural purposes. The impermeable layer which is provided for the bed and sides of canal for the purpose of improving the life and discharge capacity of canals is known as canal lining. The chapter closely examines the key concepts related to canals such as their designing and lining.

Main irrigation canal is the canal that delivers water to the lands to be irrigated and is composed of the off-load part (from the headworks to the head of distribution canals) and working part where distributing conducting canals branch off from the main canal.

Main canal is the major part of the conductive network, because water is delivered through it from the source to the entire irrigated territory. The main canal is composed of the head section, and off-load and working sections.

Head section is the inlet water-intake section where water is cleared from sediments if sediment tanks are available. The canal's off-load part delivers water from an irrigation source to the first distribution canal. The canal's working section distributes water between distribution canals.

The main canal is laid orienting on the highest points of a given irrigated area.

An irrigated land can be supplied with water from a source by gravity or by pump irrigation. At gravity water supply, water intake can be of dam intake and river-canal (damless) intake types. At that, water comes from a river (reservoir) through a check structure to the off-load part of the main canal by which it is transported to a particular irrigated land. At pumping irrigation, water comes through pipes.

The location and length of the main canal's off-load section depend on the slope of the ground and main canal, water withdrawal source water level rise above the earth's surface where the water needs to be taken to.

In some cases, minimum slope does not provide cost-effective solution, since the cross section becomes too large and considerably increases the volume of earthwork operations. And so the most advantageous slope is established, at which no silting will happen and the main canal construction cost will be lowest.

When laying canal, the command point to which water is required to be delivered from a river by gravity or a water withdrawal point. In the first case, it would be necessary to determine the length of the main canal's off-load section and a water withdrawal point. Canal laying is carried out from the command point towards the river. In the second case, it would be necessary to determine the length of the off-load section of the main canal and determine the upper boundary of the command area. The canal is laid from a water withdrawal point towards the irrigated area under design.

The main canal must command the entire irrigated area, from the end of the off-load section to the end of the last cluster of structures from which only a distribution network is constructed.

Depending on the relief, one can outline three key layouts of the main canal and inter-farm network:

The main canal is routed towards the horizontals (across the slope) at a design gradient along the upper boundary of the respective irrigated area. First-order distributors are designed as double-sided command type along the maximum slope. If there are hollows, uplands and subdivides, the main canal is designed based on the same principle; however, along the upper boundary it loses its straightness. Distribution canals are routed at an allowable gradient along the dividing crests and to ensure commanding both slopes.

The main canal is laid along the maximum surface slope, approximately in the middle of the designed irrigated land. The distributors are laid at minimum allowable gradient along the land contours. Such a layout of canals is reasonable to apply for the irrigated lands with long length along the slope (gradient) and narrow width. In this case, drop structures are designed only on the main canal, while in the first layout version those have to be designed on all first-order distributors. At considerable flow rate in the main canal, a hydropower plant is constructed in the places of concentrated water fall.

Under the conditions of complex relief, the main canal, inter-farm, and farm distributors are arranged on command points and watershed divides so that the flow velocities through canals should be lower than the scouring (erosive) velocity and higher than the silting velocity and the requirements to the management of such an irrigated area as well as other conditions of canal designing should be met.

Uses and Effects of Canal Irrigation

- Cheap labour and availability of cement reduces the cost of canal construction.

- Huge quantities of water from Monsoon rainfall & melting of snow can be stored in reservoirs during summer season.

- Irregular supply of water in the rivers is then regulated by construction of dams & barrages. Canal system irrigates a vast area. Even the deserts have been made productive.

Causes:

- Abundance of silt eroded from the Karakoram, Hindu Kush and Himalayan mountains.

- Deforestation - ruthless cutting of trees for fuel and timber. Rivers form narrow and deep valleys in the mountainous areas. Most of the eroded material is washed down into the plains and piles up in reservoirs of the dams.

Effects:

- Blockage of canals because silt accumulates.

- Weakens the foundation of dams.

- Reduced capacity of reservoir and less flow of water affects the generation of hydro-electric power. It also results in availability of less water for irrigation purposes.

- Flow of floodwater is hampered which may cause heavy damage to the dam because of mounds of silt which block the flow of water.

- Large-scale afforestation especially on the foothills of Himalayas.

- Cemented embankment of canals.

- Installation of silt trap before the water flows to the dams.

- Structural measures such as operating the reservoir at lower level during flood and allowing free flow during low flow season for sluicing sediments from the reservoir.

Uses of Irrigation:

- Soft soil and level land of the Indus Plain makes digging of canals easier than in the rugged lands of Balochistan.

- By canal irrigation millions of gallons of water are utilized that would flow into the Arabian Sea.

- Cheap labor and availability of cement reduces the cost of canal construction.

- Canal system irrigates a vast area. Even the deserts have been made productive.

- Irregular supply of water in the rivers is then regulated by construction of dams & barrages.

- Huge quantities of water from Monsoon rainfall & melting of snow can be stored in reservoirs during summer season.

- Southward slope of the rivers makes construction of canals easier, because water flows southwards naturally.

Canal Lining

Canal Linings are provided in canals to resist the flow of water through its bed and sides. These can be constructed using different materials such as compacted earth, cement, concrete, plastics, boulders, bricks etc. The main advantage of canal lining is to protect the water from seepage loss.

Canal Lining.

Canal Lining is an impermeable layer provided for the bed and sides of canal to improve the life and discharge capacity of canal. 60 to 80% of water lost through seepage in an unlined canal can be saved by construction canal lining.

Types of Canal Linings

Canal linings are classified into two major types based on the nature of surface and they are:

- Earthen type lining.

- Hard surface lining.

Earthen Type Lining

Earthen Type lining are again classified into two types and they are as follows;

- Compacted Earth Lining.

- Soil Cement Lining.

Compacted Earth Lining

Compacted earth linings are preferred for the canals when the earth is available near the site of construction or In-situ. If the earth is not available near the site then it becomes costlier to construct compacted earth lining.

Compaction reduces soil pore sizes by displacing air and water. Reduction in void size increases the density, compressive strength and shear strength of the soil and reduces permeability. This is accompanied by a reduction in volume and settlement of the surface. Proper compaction is essential to increase the stability and frost resistance (where required) and to decrease erosion and seepage losses.

Compacted Earth Lining.

Soil Cement Lining

Soil-cement linings are constructed with mixtures of sandy soil, cement and water, which harden to a concrete-like material. The cement content should be minimum 2-8% of the soil by volume. However, larger cement contents are also used.

In general, for the construction of soil-cement linings following two methods are used:

- Dry-mix method,

- Plastic mix method.

For erosion protection and additional strength in large channels, the layer of soil-cement is some-times covered with coarse soil. It is recommended the soil-cement lining should be protected from the weather for seven days by spreading approximately 50 mm of soil, straw or hessian bags over it and keeping the cover moistened to allow proper curing. Water sprinkling should continue for 28 days following installation.

Soil Cement Lining.

Hard Surface Canal Linings

It is sub divided into 4 types and they are:

- Cement Concrete Lining,

- Brick Lining,

- Plastic Lining,

- Boulder Lining.

Cement Concrete Lining

Cement Concrete linings are widely used, with benefits justifying their relatively high cost. They are tough, durable, relatively impermeable and hydraulically efficient. Concrete linings are suit-able for both small and large channels and both high and low flow velocities. They fulfill every purpose of lining.

There are several procedures of lining using cement concrete:

- Cast in situ lining.

- Shortcrete lining.

- Precast concrete lining.

- Cement mortar lining.

Cement Concrete Lining.

Brick Lining

In case of brick lining, bricks are laid using cement mortar on the sides and bed of the canal. After laying bricks, smooth finish is provided on the surface using cement mortar.

Construction of Brick Canal Lining.

Plastic Lining

Plastic lining of canal is newly developed technique and holds good promise. There are three types of plastic membranes which are used for canal lining, namely:

- Low density poly ethylene.
- High molecular high density polythene.
- Polyvinyl chloride.

The advantages of providing plastic lining to the canal are many as plastic is negligible in weight, easy for handling, spreading and transport, immune to chemical action and speedy construction.

The plastic film is spread on the prepared sub-grade of the canal. To anchor the membrane on the banks 'V trenches are provided. The film is then covered with protective soil cover.

Boulder Lining

This type of lining is constructed with dressed stone blocks laid in mortar. Properly dressed stones are not available in nature. Irregular stone blocks are dressed and chipped off as per requirement.

When roughly dressed stones are used for lining, the surface is rendered rough which may put lot of resistance to flow. Technically the coefficient of rugosity will be higher. Thus the stone lining is limited to the situation where loss of head is not an important consideration and where stones are available at moderate cost.

Advantages of Canal Lining

1. Seepage Reduction.

2. Prevention of Water Logging.

3. Increase in Commanded Area.

4. Increase in Channel Capacity.

5. Less Maintenance.

6. Safety Against Floods.

1. Seepage Reduction

The main purpose behind the lining of canal is to reduce the seepage losses. In some soils, the seepage loss of water in unlined canals is about 25 to 50% of total water supplied. The cost of canal lining is high but it is justifiable for its efforts in saving of most of the water from seepage losses. Canal lining is not necessary if seepage losses are very small.

2. Prevention of Water Logging

Water logging is caused due to phenomenal rise in water table due to uncontrolled seepage in an unlined canal. This seepage effects the surrounding ground water table and makes the land unsuitable for irrigation. So, this problem of water logging can be surely prevented by providing proper lining to the canal sides.

3. Increase in Commanded Area

Commanded area is the area which is suitable for irrigation purpose. The water carrying capacity of lined canal is much higher than the unlined canal and hence more area can be irrigated using lined canals.

4. Increase in Channel Capacity

Canal lining can also increase the channel capacity. The lined canal surface is generally smooth and allows water to flow with high velocity compared to unlined channel. Higher the velocity of flow greater is the capacity of channel and hence channel capacity will increase by providing lining.

On the other side with this increase in capacity, channel dimensions can also be reduce to maintain the previous capacity of unlined canal which saves the cost of the project.

5. Less Maintenance

Maintenance of lined canal is easier than unlined canals. Generally there is a problem of silting in unlined canal which removal requires huge expenditure but in case of lined canals, because of high velocity of flow, the silt is easily carried away by the water.

In case of unlined canals, there is a chance of growth of vegetation on the canal surface but not in case of lined canals. The vegetation affect the velocity of flow and water carrying capacity of channel. Lined canal also prevents damage of canal surface due to rats or insects.

6. Safety against Floods

A line canal always withstand against floods while unlined canal may not resists and also there is chance of occurring of breach which damages the whole canal as well as surrounding areas or fields. But among the all concrete canal linings are good against floods or high velocity flows.

Design of Irrigation Canals

The entire water conveyance system for irrigation, comprising of the main canal, branch canals, major and minor distributaries, field channels and water courses have to be properly designed. The design process comprises of finding out the longitudinal slope of the channels and fixing the cross sections. The channels themselves may be made up of different construction materials. For example, the main and branch canals may be lined and the smaller ones unlined. Even for the unlined canals, there could be some passing through soils which are erodible due to high water velocity, while some others may pass through stiff soils or rock, which may be relatively less prone to erosion. Further, the bank slopes of canals would be different for canals passing through loose or stiff soils or rock. We discuss the general procedures for designing canal sections, based on different practical considerations.

The earlier practice had been to provide triangular channel sections with rounded bottom for smaller discharges. The geometric elements of these two types of channels are given below:

Triangular Section

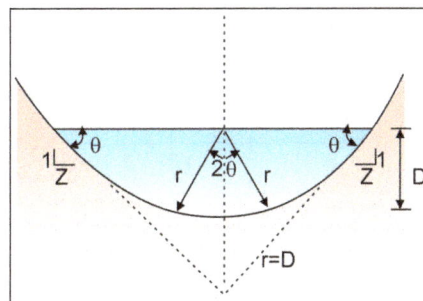

Triangular channel section.

For triangular section, the following expressions may be derived

$\quad$ A = D² (..cot)

$\quad$ P = 2 D (..cot)

$\quad$ R = D / 2.

The above expressions for cross sectional area (A), wetted perimeter (P) and hydraulic radius (R) for a triangular section may be verified by the reader.

Trapezoidal Section

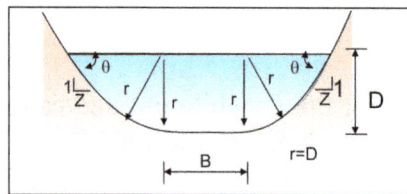

Trapezoidal channel section.

For the Trapezoidal channel section, the corresponding expressions are:

$\quad$ A = B D + D² (.. cot)

$\quad$ P = B + 2 D (..cot)

$\quad$ R = A / P

The expressions for A and P may, again, be verified by the reader. In all the above expressions, the value of is in radians.

The steps to be followed for selecting appropriate design parameters of a lined irrigation channel, according to IS: 10430 may be summarized as follows:

1. Select a suitable slope for the channel banks. These should be nearly equal to the angle of repose of the natural soil in the subgrade so that no earth pressure is exerted from behind on the lining. For example, for canals passing through sandy soil, the slope may be kept as 2H: 1V whereas canals in firm clay may have bank slopes as 1.5H: 1V canals cut in rock may have almost vertical slopes, but slopes like 0.25 to 0.75H: 1V is preferred from practical considerations.

2. Decide on the freeboard, which is the depth allowance by which the banks are raised above the full supply level (FSL) of a canal. For channels of different discharge carrying capacities, the values recommended for freeboard are given in the following table:

Type of Channel	Discharge (m³/s)	Freeboard (m)
Main and branch canals	> 10	0.75
Branch canals and major distributaries Major	5 – 10	0.6
distributaries	1 – 5	0.50
Minor distributaries	< 1	0.30
Water courses	< 0.06	0.1 – 0.15

3. Berms or horizontal strips of land provided at canal banks in deep cutting, have to be incorporated in the section.

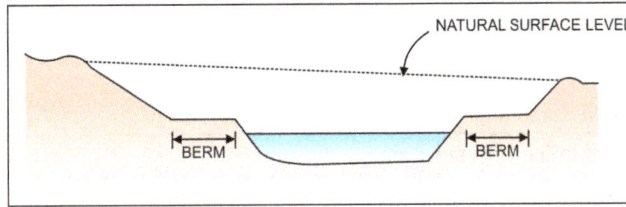

The berms serve as a road for inspection vehicles and also help to absorb any soil or rock that may drop from the cut-face of soil or rock of the excavations. Berm width may be kept at least 2m. If vehicles are required to move, then a width of at least 5m may be provided.

4. For canal sections in filling, banks on either side have to be provided with sufficient top width for movement of men or vehicles.

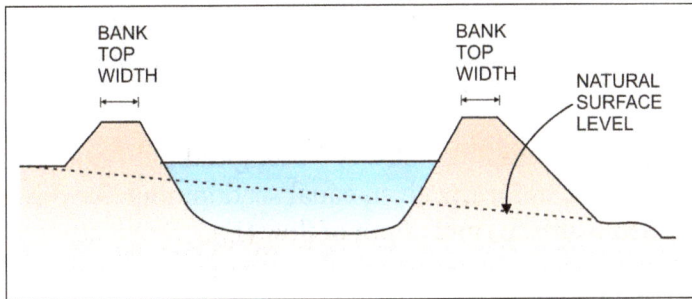

Canal sections in filling.

The general recommendations for bank top width are as follows:

Discharge (m^3/s)	Maximum bank top width (m)	
	For inspection road	For non-inspection banks
0.15 to 7.5	5.0	1.5
7.5 to 10.0	5.0	2.5
10.0 to 15.0	6.0	2.5
15.0 to 30.0	7.0	3.5
Greater than 30.0	8.0	5.0

Next, the cross section is to be determined for the channel section.

5. Assume a safe limiting velocity of flow, depending on the type of lining, as given below:

- Cement concrete lining: 2.7 m/s.

- Brick tile lining or burnt tile lining: 1.8 m/s.

- Boulder lining: 1.5 m/s.

6. Assume the appropriate values of flow friction coefficients. Since Manning's equation would usually be used for calculating the discharge in canals, values of Manning's roughness coefficient, n, from the following table may be considered for the corresponding type of canal lining.

Surface Characteristics		Value of n
Concrete with surfaces as:		
a)	Formed, no finish/PCC tiles or slabs	0.018-0.02
b)	Trowel float finish	0.015-0.018
c)	Gunited finish	0.018-0.022
Concrete bed trowel finish with sides as:		0.019-0.021
a)	Hammer dressed stone masonry	
b)	Course rubble masonry	0.018-0.02
c)	Random rubble masonry	0.02-0.025
d)	Masonry plastered	0.015-0.017
e)	Dry boulder lining	0.02-0.03
Brick tile lining		0.018-0.02

7. The longitudinal slope (S) of the canal may vary from reach to reach, depending upon the alignment. The slope of each reach has to be evaluated from the alignment of the canal drawn on the map of the region.

8. For the given discharge Q, permissible velocity V, longitudinal slope S, given side slope , and Manning' roughness coefficient, n, for the given canal section, find out the cross section parameters of the canal, that is, bed width (B) and depth of flow (D).

Since two unknowns are to be found, two equations may be used, which are:

- Continuity equation: Q = A * V.

- Dynamic equation: $V = \dfrac{1}{n}\left(A\,R^{2/3}\,S^{1/2}\right)$.

In the above equations, all variables stand for their usual notation as mentioned earlier, A and R is cross sectional area and hydraulic radius, respectively

Typical Sections of Lined Channels

Though there may be a large number of combinations of the factors on which the cross-section of a lined canal depends, some typical examples are given in the following figures, which may give an idea of laying and a practical channel cross section.

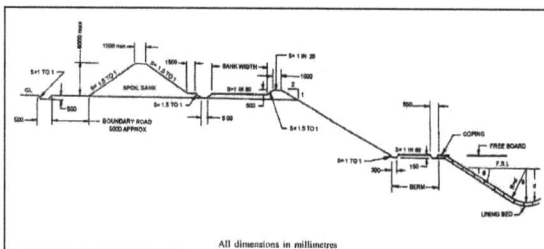

(a) Typical cross section of canals when natural ground level is below bed level.

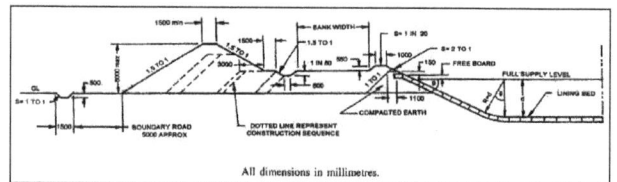

(b) Typical cross section of canals when natural ground level is between canal and full supply levels.

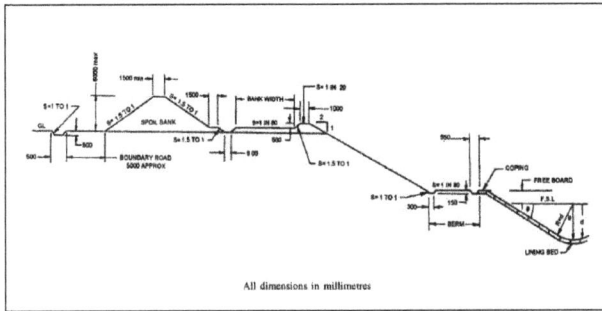

(c) Typical cross section of canals when natural ground level is above the top level of lining.

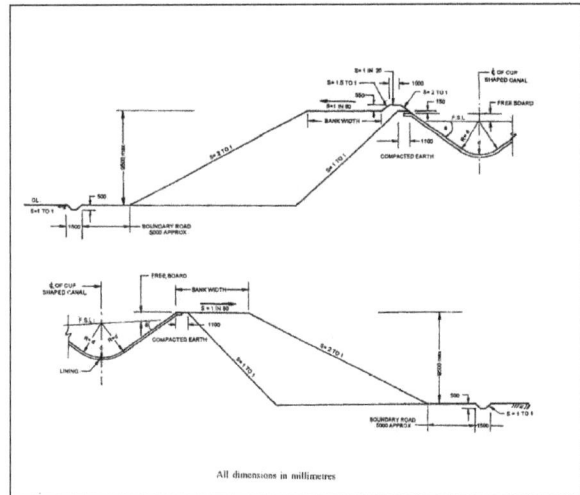

(d) Typical cross section of canals when canal bed level is above natural ground level.

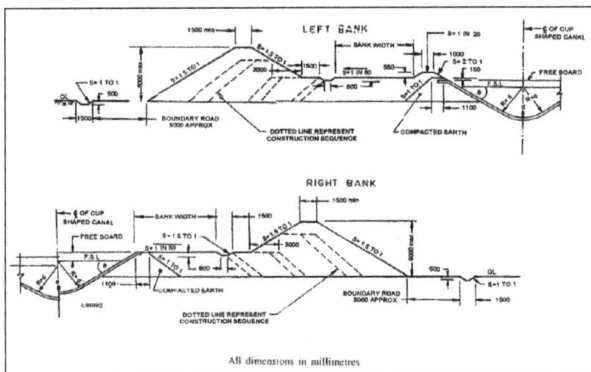

(e) Typical cross section of canals when natural ground level is between canal bed and full supply levels.

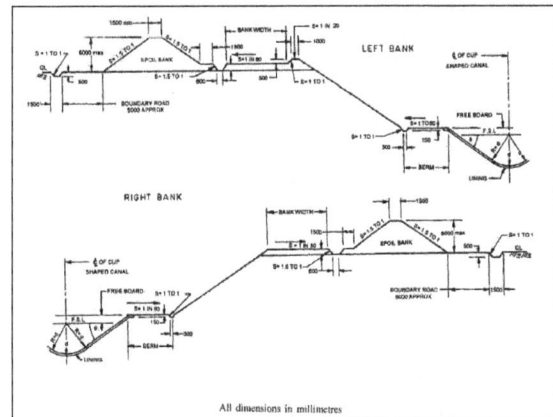

(f) Typical cross section of canals when natural ground level is above top of canal lining.

Subsurface Drainage of Lined Canals

Lined canals passing through excavations may face a situation when the canal is dry and the surrounding soil is saturated, like when the ground table is very near the surface. Similar situation may occur for lined canals in filling when the confining banks become saturated, as during rains and the canal is empty under the circumstances of repair of lining or general closure of canal. The hydrostatic pressure built up behind the linings, unless released, causes heaving of the lining material, unless it is porous enough to release the pressure on its own. Hence, for most of the linings (except for the porous types like the boulder or various types of earth linings which develop inherent cracks), there is a need to provide a mechanism to release the back pressure of the water in the subgrade. This may be done by providing pressure relief valves.

Code of practice for under design of lined canals" (First revision) discusses various methods for relieving uplift pressure below canal linings.

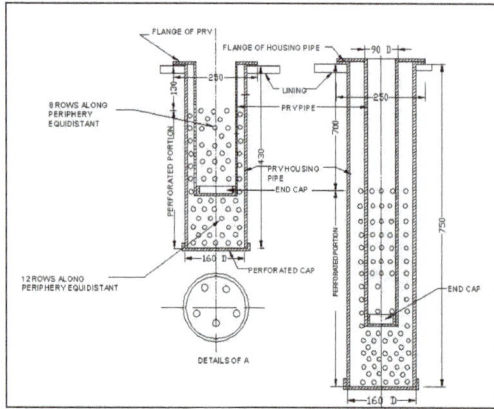

(a) Details of a Pressure Relief Valve (PRV).

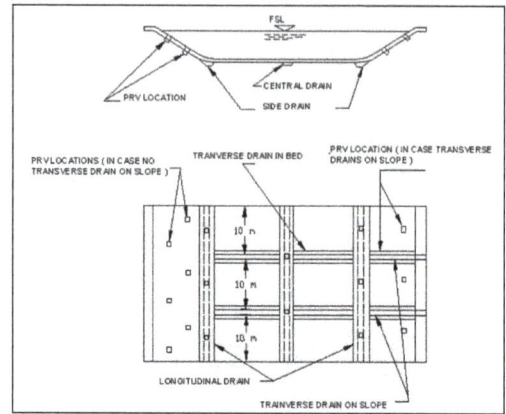

(b) Possible locations of PRVs.

Design of Unlined Canals

The Bureau of Indian Standard code IS: 7112-1973 "Criterion for design of crosssection for unlined canals in alluvial soils" is an important document that may be consulted for choosing various parameters of an unlined channel, specifically in alluvial soils. There are unlined canals flowing through other types of natural material like silty clay, but formal guidelines are yet to be brought out on their design. Nevertheless, the general principles of design of unlined canals in alluvial soils are enumerated here, which may be suitably extended for other types as well after analyzing prototype data from a few such canals.

The design of unlined alluvial canals as compared to lined canals is more complex since here the bed slope cannot be determined only on the basis of canal layout, since there would be a limiting slope, more than which the velocity of the flowing water would start eroding the particles of the canal bed as well as banks. The problem becomes further complicated if the water entering the canal from the head-works is itself carrying sediment particles. In that case, there would be a limiting slope, less than which the sediment particles would start depositing on the bed and banks of the canal. In the following sections the design concept of unlined canals in alluvium for clear water as well as sediment-laden water is discussed separately.

Unlined Alluvial Canals in Clear Water

A method of design of stable channels in coarse non-cohesive material carrying clear water has been developed by the United States Bureau of Reclamation as reported by Lane (1955), which is commonly known as the Tractive Force Method. Figure shows schematically shows such a situation where the banks are inclined to the horizontal at a given angle θ.

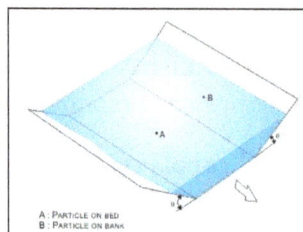

Trapezoidal shaped unlined alluvial canal Two particles A and B are on the bed and bank.

It is also assumed that the particles A and B both have the same physical properties, like size, density, etc. and also possess the same internal friction angle Φ. Naturally, the bank inclination θ should be less than Φ, for the particle B to remain stable, even under a dry canal condition. When there is a flow of water, there is a tendency for the particle A to be dragged along the direction of canal bed slope, whereas the particle B tries to get dislodged in an inclined direction due to the shear stress of the flowing water.

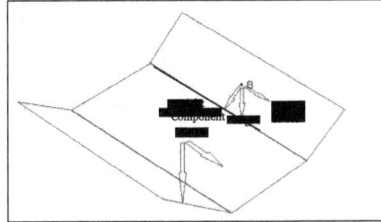

Shear stress and weight components of the particles A and B on the bed and bank, respectively.

The particle A would get dislodged when the shear stress, τ, is just able to overcome the frictional resistance. This critical value of shear stress is designated as τ_c may be related to the weight of the particle, W, as:

$$\tau_{Cb} = W \tan \phi$$

For the particle B, a smaller shear stress is likely to get it dislodged, since it is an inclined plane. In fact, the resultant of its weight component down the plane, W Sin , and the shear stress (designated as $'\tau_c$) would together cause the particle to move. Hence, in this case,

$$\left(\tau_{CS}\right)^2 + \left(W \sin \theta\right)^2 = \left(W \cos \theta\right) \tan \phi$$

In the above expression it must be noted, that the normal reaction on the plane for the particle B is W Cos θ.

Eliminating the weight of the particles, W, from both above equations, one obtains,

$$\tau_{CS}^2 + \left[\frac{\tau_{Cb}}{\tan \phi} \sin \theta\right]^2 = \left[\frac{\tau_{Cb}}{\tan \phi} \cos \theta \tan \phi\right]^2$$

This simplifies to,

$$\tau_{CS}^2 = \tau_{Cb}^2 \left[\cos^2 \theta - \frac{\sin^2 \theta}{\tan^2 \phi}\right]$$

Or

$$\frac{\tau_{CS}}{\tau_{Cb}} = \cos \theta \sqrt{1 - \frac{\tan^2 \theta}{\tan^2 \phi}}$$

As expected, τ_{cs} is less than τ_{cb}, since the right hand side expression of equation (R = D / 2) is less than 1.0. This means that the shear stress required moving a grain on the side slope is less than that required to move on the bed. It is now required to find out an expression for the shear stress due to flowing water in a trapezoidal channel. It is known that in a wide rectangular channel, the shear stress at the bottom, τ_0 is given by the following expression.

$$\tau_0 = \gamma\, R\, S$$

Where γ is the unit weight of water, R is the hydraulic radius of the channel section and S is the longitudinal bed slope. Actually, this is only an average value of the shear stress acting on the bed, but actually, the shear stress varies across the channel width. Studies conducted to find the variation of shear stress have revealed interesting results, like the variation of maximum shear stress at channel base (τ_b) and sides (τ_s)

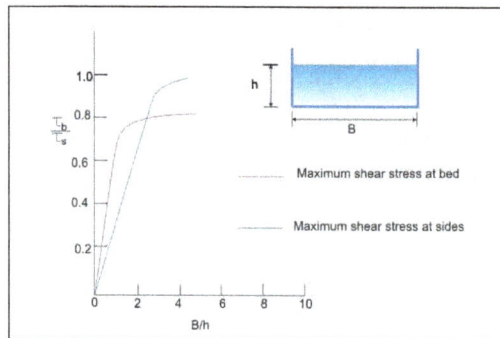

Variation of maximum shear stress for rectangular channel.

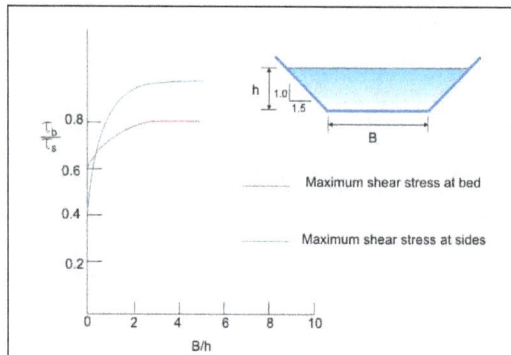

Variation of maximum shear stress of trapezoidal channel with side slope 1V:1.5H.

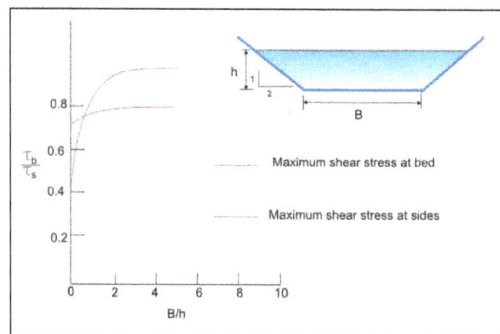

Variation of maximum shear stress of trapezoidal channel with side slope 1V :2H.

As may be seen from the above figures, for any type of channel section, the maximum shear stress at the bed is somewhat more than for that at the sides for a given depth of water (Compare τ_b and

τ_s for same B/h value for any graph). Very roughly, for trapezoidal channels with a wide base compared to the depth as is practically provided, the bottom stress may be taken as γRS and that at the sides as $0.75\ \gamma RS$. Finally, it remains to find out the values of B and h for a given discharge Q that may be passed through an unlined trapezoidal channel of given side slope and soil, such that both the bed and banks particles are dislodged at about the same time. This would ensure an optimum channel section.

Researchers have investigated for long, the relation between shear stress and incipient motion of non-cohesive alluvial particles in the bed of a flowing stream. One of the most commonly used relation, as suggested by Shields.

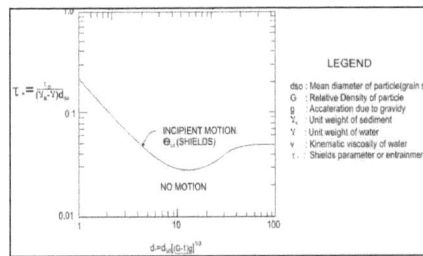

Curve for incipient motion.

Swamee and Mittal (1976) have proposed a general relation for the incipient motion which is accurate to within 5 percent. For $\gamma s = 2650$ kg/m³ and $\gamma = 1000$ kg/ m³ the relation between the critical shear stress τ_c (in N/m²) diameter of particle d_s (in mm) is given by the equation:

$$\tau_C = 0.155 + \frac{0.409\ d_s^2}{\sqrt{1+0.177 d_s^2}}$$

The application of the above formula for design of the section may be illustrated with an example.

Say, a small trapezoidal canal with side slope 2H: 1V is to be designed in a soil having an internal friction angle of 35° and grain size 2mm. The canal has to be designed to carry 10m³/s on a bed slope of 1 in 5000.

To start with, we find out the critical shear stress for the bed and banks. We may use the graph in figure or; more conveniently, use Equation ($\tau_C = 0.155 + \frac{0.409\ d_s^2}{\sqrt{1+0.177 d_s^2}}$). Thus, we have the critical shear stress for bed, τ_{Cb}, for bed particle size of 2mm as:

$$\tau_C = \tau_{Cb} = 0.155 + \frac{0.409*2^2}{\sqrt{1+0.177*2^2}}$$

$$= 1.407 \text{ N/m}^2$$

The critical shear stress for the sloping banks of the canal can be found out with the help of expression ($\frac{\tau_{CS}}{\tau_{Cb}} = \cos\theta\sqrt{1-\frac{\tan^2\theta}{\tan^2\phi}}$). Using the slope of the banks (2H: 1V), which converts to = 26.6°

$$\frac{\tau_{CS}}{\tau_{Cb}} = \cos 26.6^0\sqrt{1-\frac{\tan^2 26.6^0}{\tan^2 35^0}} = 0.625$$

From which,

$$\tau_{CS} = 0.625 \times 0.1.4068 = 0.880 \text{ N/m}^2$$

The values for the critical stresses at bed and at sides are the limiting values. One does not wish to design the canal velocity and water depth in such a way that the actual shear stress reaches these values exactly since a slight variation may cause scouring of the bed and banks. Hence, we adopt a slightly lower value for each, as:

Allowable critical shear stress for bed $\tau'_{Cb} = 9.0 \ \tau_{Cb} = 1.266 \text{ N/m}^2$

Allowable critical shear stress for banks $\tau'_{CS} = 9.0 \ \tau_{CS} = 0.792 \text{ N/m}^2$

The dimensions of the canal is now to be determined, which means finding out the water depth D and canal bottom B. for this, we have to assume a B/D ratio and a value of 10 may be chosen for convenience. We now read the shear stress values of the bed and banks in terms of flow variable 'R', the hydraulic radius, canal slope 'S' and unit weight of water γ from the figure-corresponding to a channel having side slope 2H: 1V. However, approximately we may consider the bed and bank shear stresses to be γ RS and 0.75 γ RS, respectively. Further, since we have assumed a rather large value of B/D, we may assume R to be nearly equal to D. this gives the following expressions for shear stresses at bed and bank;

Unit shear stress at bed $= \tau_b = \gamma \, D \, S = 9810 \times D \times 5000 \ 1 = 1.962 \ D \ \text{N/m}^2$ per metre width.

Unit shear stress at bank $\tau_s = 0.75 \ \gamma \, D \, S = 0.75 \times 9810 \times D \times 5000 \ 1 = 1.471 \ D \ \text{N/m}^2$ per metre width.

For stability, the shear stresses do not exceed corresponding allowable critical stresses.

Thus,

$$\left(\tau_b = \right) 1.962 \, D < \left(\tau_{Cb} = \right) 1.266 \, N / m^2$$
$$or \ D < 0.645 \, m$$
$$and$$
$$\left(\tau_s = \right) 1.471 D < \left(\tau_{CS} = \right) 0.792 N / m^2$$
$$or \ D < 0.538 \, m$$

Therefore, the value of D satisfying both the expression is the minimum value of the two, which means D should be limited to 0.538 m, say 0.53 m. Since the B/D ratio was chosen to be 10, we may assume B to be 5.3 m, or, say, 5.5 m for practical purposes. For a trapezoidal shaped channel with side slopes 2H: 1V, we have:

$$A = D(B + 2D) = 3.445 \, m^2$$

And

$$P = B + 2\sqrt{5} \, D = 7.87 \, m$$

$$\text{Thus } R = A / P = \frac{D(B + 2D)}{B + 2v5D} = 0.438m$$

For the grain size 2mm, we may find the corresponding Manning's roughness coefficient 'n' using the Stricker's formula given by the expression:

$$n = \frac{d_s^{1/6}}{25.6}$$

$$= \frac{0.002^{1/6}}{25.6} = 0.014$$

Using the Manning's equation of flow, we have:

$$Q = \frac{1}{n} A R^{2/3} S^{1/2}$$

$$= \frac{1}{0.014} \times 3.445 \times 0.438^{2/3} \times \left(\frac{1}{5000}\right)^{1/2}$$

$$= 2 \ m^3 / s$$

Since the value of Q does not match the desired discharge that is to be passed in the channel, given in the problem as 100m³ /s, we have to change the B/D ratio, which was assumed to be 10. Suppose we assume a B/D ratio of, say, k we obtain the following expression for the flow:

$$Q = \frac{1}{n} A \left(\frac{A}{P}\right)^{2/3} S^{1/2}$$

$$= \frac{1}{n} \frac{A^{5/3}}{P^{2/3}} S^{1/2}$$

And substituting known values, we obtain:

$$10 = \frac{1}{0.014} \times \frac{\left[D(k.D + 2D)\right]^{5/3}}{\left[k.D + 2\sqrt{5}D\right]^{2/3}} \times \left(\frac{1}{5000}\right)^{1/2}$$

Substituting the value of D as 0.53m, as found earlier, it remains to find out the value of k from the above expression. It may be verified that the value of k is evaluates to around 55, from which the bed width of the canal, B, is found out to be 29.15m, say, 30m, for practical purposes.

It may be noted that IS: 7112-1973 gives a list of Manning's n values for different materials. However, it recommends that for small canals $\left(Q < 15m^3 / S\right)$ n may be taken as 0.02. (In the above example, n was evaluated as 0.014 by Strickler's formula).

It is natural for channel carrying sediment particles along with its flow to deposit them if the velocity is slower than a certain value. Velocity in excess of another limit may start scouring the bed and banks. Hence, for channels carrying a certain amount of sediment may neither deposit, nor scour for a particular velocity.

'Criteria for design of cross section for unlined canals in alluvial soils", which prescribes that the following equations have to be used:

$$S = \frac{0.003\, f^{5/3}}{Q^{1/6}}$$

$$P = 4.75\sqrt{Q}$$

$$R = 0.47\left(\frac{Q}{f}\right)^{1/3}$$

Where the variables are as explained below:

- S: Bed slope of the channel.

- Q: the discharge in m3/s.

- P: wetted perimeter of the channel, in m.

- R: Hydraulic mean radius, in m.

- f: The silt factor for the bed particles, which may be found out by the following formula, in which d^{50} is the mean particle size in mm.

  $$f = 1.76\sqrt{d_{50}}\;.$$

It may be noted that the regime equations proposed by Lacey are actually meant for channels with sediment of approximately 500ppm. Hence, for canals with other sediment loads, the formula may not yield correct results, as has been pointed out by Lane, Blench and King, Simons and Alberts on, etc. however, the regime equations proposed by Lacey are used widely in India, though it is advised that the validity of the equations for a particular region may be checked before applying the same. For example, Lacey's equations have been derived for non-cohesive alluvial channels and hence very satisfactory results may not be expected from lower reaches of river systems where silty or silty-clay type of bed materials are encountered, which are cohesive in nature.

Application of Lacey's regime equations generally involves problems where the discharge (Q), silt factor (f) and canal side slopes (Z) are given and parameters like water depth (D), canal bed width (B) or canal longitudinal slope (S) have to be determined or Conversely, if S is known for a given f and Z, it may be required to find out B, D and Q.

Longitudinal Section of Canals

The cross section of an irrigation canal for both lined and unlined cases was discussed in the previous sections. The longitudinal slope of a canal therefore is also known or is adopted with reference to the available country slope. However, the slope of canal bed would generally be constant along certain distances, whereas the local ground slope may not be the same. Further in the alignment of a canal system was shown to be dependent on the topography of the land and other factors. The next step is to decide on the elevation of the bed levels of the canal at certain intervals along its route, which would allow the field engineers to start canal construction at the exact locations.

Also, the full supply level (FSL) of the canal has to be fixed along its length, which would allow the determination of the bank levels.

The exercise is started by plotting the plan of the alignment of the canal on a ground contour map of the area plotted to a scale of 1 in 15,000,. At each point in plan, the chainages and bed elevations marked clearly. The canal bed elevations and the FSLs at key locations (like bends, divisions, etc) are marked on the plan. It must be noted that the stretches AB and BC of the canal shall be designed that different discharges due to the offtaking major distributary. Hence, the canal bed slope could be different in the different stretches.

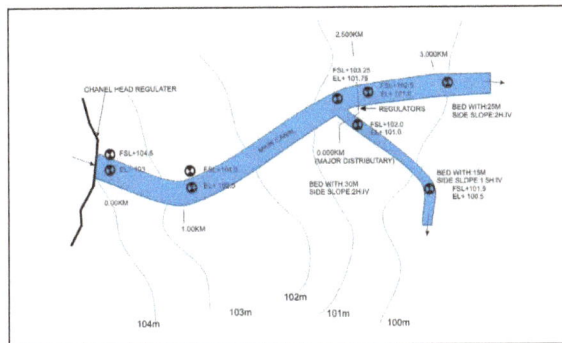

Typical layout of a canal showing bed and canal full supply levels.

The determination of the FSL starts by calculating from the canal intake, where the FSL is about 1m below the pond level on the upstream of the canal head works. This is generally done to provide for the head loss at the regulator as the water passes below the gate. It is also kept to maintain the flow at almost at full supply level even if the bed is silted up to some extent in its head reaches. On knowing the FSL and the water supply depth, the canal bed level elevation is fixed at chainage 0.00KM, since this is the starting point of the canal. At every key location, the canal bed level is determined from the longitudinal slope of the canal, and is marked on the map. If there is no offtake between two successive key locations and no change in longitudinal slope is provided, then the crosssection would not be changed, generally, and accordingly these are marked by the canal layout.

At the offtakes, where a major or minor distributary branches off from the main canal, there would usually be two regulators. One of these, called the cross regulator and located on the main canal heads up the water to the desired level such that a regulated quantity of water may be passed through the other, the head regulator of the distributary by controlling the gate opening. Changing of the cross regulator gate opening has to be done simultaneously with the adjustment of the head regulator gates to allow the desired quantity of water to flow through the distributary and the remaining is passed down the main canal.

The locus of the full supply levels may be termed as the full supply line and this should generally kept above the natural ground surface line for most of its length such that most of the commanded area may be irrigated by gravity flow. When a canal along a watershed, the ground level on its either side would be sloping downward, and hence, the full supply line may not be much above the ground in that case. In stretches of canals where there is no offtake, the canal may run through a cutting within an elevated ground, and in such a case, the full supply line would be lower than the average surrounding ground level. In case irrigation is proposed for certain reaches of the canal where the adjacent ground level is higher than the supply level of the canal, lift irrigation by pumping may be adapted locally for the region.

Similarly, for certain stretches of the canal, it may run through locally low terrain. Here, the canal should be made on filling with appropriate drainage arrangement to allow the natural drainage water to flow below the canal. The canal would be passing over a water-carrying bridge, called aqueducts, in such a case.

As far as possible, the channel should be kept in balanced depth of cutting and filling for greatest economy and minimum necessity of borrow pits and spoil banks.

The desired canal slope may, at times, is found to be much less than the local terrain slope. In such a case, if the canal proceeds for a long distance, an enormous amount of filling would be required. Hence, in such a case, canal falls are provided where a change in bed elevation is effected by providing a drop structure usually an energy dissipater like hydraulic jump basin is provided to kill the excess energy gained by the fall in water elevation. At times, the drop in head is utilized to generate electricity through suitable arrangement like a bye-pass channel installed with a bulb-turbine.

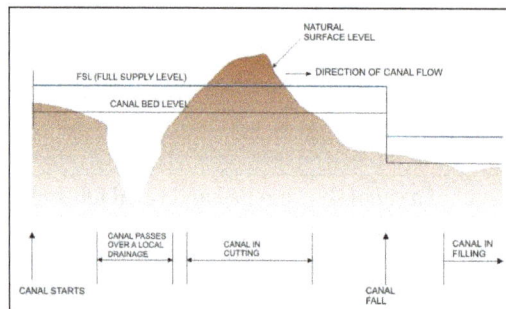

Longitudinal section of a canal assuming no withdrawals in this stretch.

A typical canal section is shown in figure, for a canal stretch passing through varying terrain profile. Here, no withdrawals have been assumed and hence, the discharge in the entire stretch of the canal is assumed to remain same. Hence, the canal bed slope and water depth are also not shown varying. It is natural that if the canal has outlets in between, the change in discharge would result in corresponding changes in the full supply line.

Minimum Cost Design of Lined Canal Sections

Lining of a canal is essential for efficient use of land and water resources. Control of seepage saves water for further extension of the irrigation network as well as reduces the water logging in the adjoining areas. The smooth surface of lining reduces the friction slope, which enables the canal to be laid on a flatter bed slope. This increases the command area of the canal. On the other hand, as the lining permits higher average velocities, the canal can be laid on steeper slopes to save the cost of earthwork in formation. As the lining provides a rigid boundary, it ensures protection against bed and bank erosion. When the canal is constructed in an area containing black cotton soil, the canal lining becomes essential for stability of the canal banks. Thus, the maintenance cost gets reduced.

Canals conveying water in arid and semi arid regions need to be lined prior to conveying water in them. This is because, for a newly completed irrigation project, seepage from the canal is maximum as the ground water level is likely to be at a larger depth below the canal bed. Also in arid and semi arid regions the ground water is likely to be brackish and the seepage water which joins the ground water may not be withdrawn by pumping as the pumped water is unlikely to satisfy the irrigation water standards. Seepage control for existing canals with conventional lining is found to be prohibitive because of material costs and restriction on closure of the canal. If lining is envisaged in the planning stage, a smaller cross section could be adopted and lining can be justified from an economic point of view.

Several types of materials are used for canal lining. The choice of material mainly depends on the degree of water tightness required. Though less watertight, soil-cement lining and boulder lining are preferred on account of their low initial cost. Another low cost lining is composed of polyethylene plastic or alkathene sheets spread over the boundary surface with adequate earth cover. This type of lining is used for stable channels. Brick lining and brunt clay tile lining are the popular linings as they provide reasonable water tightness along with strength. For improving these qualities, double layers of brick or the brunt clay tiles are also provided. For canals carrying large discharges, in situ concrete lining or concrete tile linings are used. Low-density polyethylene (LDPE) or high-density polyethylene (HDPE) films of thickness 100 μm, 200 μm or more are sandwiched between two layers of canal linings to act as a second line of defense.

Lined canals are designed for uniform flow formula considering hydraulic efficiency, practicability, and economy. Chow and French have listed various properties of the most hydraulically efficient sections. Swamee and Bhatia expressed all the channel dimensions in term of a length scale comprising independent design variables and developed curves for the optimal design of trapezoidal, rounded bottom and rounded corner sections. Guo and Hughes found that a channel narrower than the hydraulically best section results in minimum excavation when free board is taken into consideration. Monadjemi and Swamee have done a comprehensive investigation for the optimal dimensions for various canal shapes. Canal cost and practicability requirements are combined into objective function and constraints in obtaining the optimal channel section by several investigators. Manning's equation as a uniform flow equation was used in the optimal channel design by the above investigators. A more general resistance equation based on roughness height was used in the optimal design of irrigation canals by Swamee. Thus the review of literature reveals that though considerable work has been reported on the design of minimum area cross section, practically no work has been done on the minimum cost lined canal sections. Minimum cost design of lined canals involves minimization of the sum of depth-dependent excavation cost and cost of lining subject to uniform flow condition in the canal, which results in nonlinear objective function and nonlinear equality constraint making the problem hard to solve analytically. In this paper, generalized empirical equations for design of minimum cost lined sections are obtained for triangular, rectangular, trapezoidal, and circular canals.

Resistance

Equation Uniform open channel flow is governed by the resistance equation. The most commonly used resistance formula is Manning's equation which is applicable for rough turbulent flow, and in a limited bandwidth of relative roughness. Relaxing these restrictions, Swamee gave the following resistance equation,

$$V = -2.457\sqrt{gRS_0}\ \ln\left(\frac{\varepsilon}{12R} + \frac{0.221v}{R\sqrt{gRS_0}}\right),$$

where V = average flow velocity (m s^{-1}); g = gravitational acceleration (m s^{-2}); R = hydraulic radius (m) defined as the ratio of the flow area A (m^2) to the flow perimeter P (m); ε = average roughness height of the canal lining (m); and v = kinematic viscosity of water (m^2 s^{-1}). Similar to the case of resistance equation for pipe flow, Equation $V = -2.457\sqrt{gRS_0}\ \ln\left(\dfrac{\varepsilon}{12R} + \dfrac{0.221v}{R\sqrt{gRS_0}}\right)$, involves physically conceivable parameters ε and v. The discharge Q (m^3 s^{-1}) was given by the continuity equation,

$$Q = AV$$

Combining Equations $V = -2.457\sqrt{gRS_0}\ \ln\left(\dfrac{\varepsilon}{12R} + \dfrac{0.221v}{R\sqrt{gRS_0}}\right)$, and $Q = AV$ the discharge was given by,

$$Q = -2.457A\sqrt{gRS_0}\ \ln\left(\frac{\varepsilon}{12R} + \frac{0.221v}{R\sqrt{gRS_0}}\right).$$

Cost Structure

The objective function consists of the cost of the canal's unit length. This includes the costs of lining and of the earthwork. Considering the unit cost of lining (cost per unit surface area covered) to be independent of the depth of placement, the cost of lining C_L (monetary unit per unit length, e.g., \$ m^{-1}) was expressed as:

$$C_L = c_L P.$$

where c_L = unit cost of lining (monetary unit per unit area of lining, e.g., \$ m-2).

Table: Lining and earthwork cost coefficients

Types of strata	cL/ce (m)									ce/cr (m)
	Type of lining									
	Concrete tile			Brick tile			Brunt clay tile			
	With LDPE film		Without film	With LDPE film		Without film	With LDPE film		Without film	
	100 μ	200 μ		100 μ	200 μ		100 μ	200 μ		
(1)	(2)	(3)	(4)	(5)	(6)	(7)	(8)	(9)	(10)	(11)
Ordinary soil	12.75	13.02	12.24	6.39	6.67	5.88	6.08	6.35	5.57	6.96
Hard soil	10.00	10.22	9.60	5.01	5.23	4.62	4.77	4.99	3.37	8.86
Impure lime nodules	8.90	9.10	8.55	4.47	4.66	4.11	4.25	4.44	3.89	9.96
Dry shoal with shingle	6.56	6.71	6.30	3.29	3.43	3.03	3.13	3.27	2.86	13.50
Slush and lahel	6.40	6.54	6.14	3.21	3.35	2.95	3.05	3.19	2.79	13.86

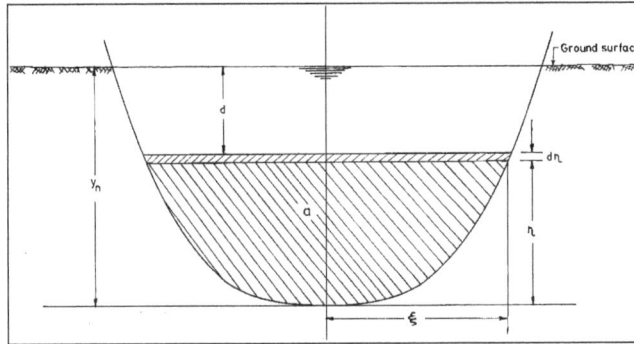

Definition sketch

For a canal section with the normal water surface at the (average) ground level, as shown in figure, the earthwork cost C_e (monetary unit per unit length, e.g., \$ m⁻¹) was written as:

$$C_e = c_e A + c_r \int_0^A (y_n - \eta)\, da,$$

where c_e = unit cost of earthwork at ground level (monetary unit per unit volume of earthwork, e.g., \$ m⁻³); c_r = the additional cost per unit excavation per unit depth (monetary unit per unit volume of excavation per unit depth, e.g. \$ m⁻⁴); η = vertical coordinate; yn = normal depth (m); a = flow area (m²) at height η. As cL/ce and c_e/c_r have length dimension, they remain unaffected by the monetary unit chosen. Using 'Schedule' and 'U.P.' the c_L/c_e and c_e/c_r ratios were obtained for various types of linings and the soil strata. Integrating by parts, Equation $C_e = c_e A + c_r \int_0^A (y_n - \eta)\, da,$ was reduced to,

$$C_e = c_e A + c_r \int_0^{y_n} a\, d\eta.$$

Adding Equations $C_L = c_L P.$ and $C_e = c_e A + c_r \int_0^{y_n} \eta\, da.$, the cost function C (monetary unit per unit length, e.g., \$ m⁻¹) was obtained as,

$$C_e = c_e A + c_r P \int_0^{y_n} a\, d\eta.$$

Optimization Algorithm

The problem of determination of optimal canal section shape was reduced to,

Minimize $C = c_e A + c_L P + c_r \int_0^{y_n} a\, d\eta.$

subject to $\phi = Q + 2.457\, A\sqrt{gRS_0}\ \ln\left(\dfrac{\varepsilon}{12R} + \dfrac{0.22\, lv}{R\sqrt{gRS_0}} \right).$

where ϕ = equality constraint function. The constrained optimization problem was solved by minimizing the augmented function ψ given by,

$$\psi = C + p\phi^2$$

where p = a penalty parameter. Adopting small p, Equation $\psi = C + p\phi^2$ was minimized using grid search algorithm. Increasing p five-fold, the minimization was carried through various cycles till the optimum stabilized.

Optimal Section Shapes

Defining the length scale λ as,

$$\lambda = \left(\frac{Q^2}{gS_0} \right)^{0.2}$$

the following nondimensional variables were obtained:

$$C_* = \frac{C}{c_e \lambda^2},$$

$$c_{L*} = \frac{c_L}{c_e \lambda}$$

$$c_{r*} = \frac{c_r \lambda}{c_e}$$

$$y_n^* = \frac{y_n}{\lambda}$$

$$\varepsilon_* = \frac{\varepsilon}{\lambda},$$

$$v_* = \frac{v\lambda}{Q}.$$

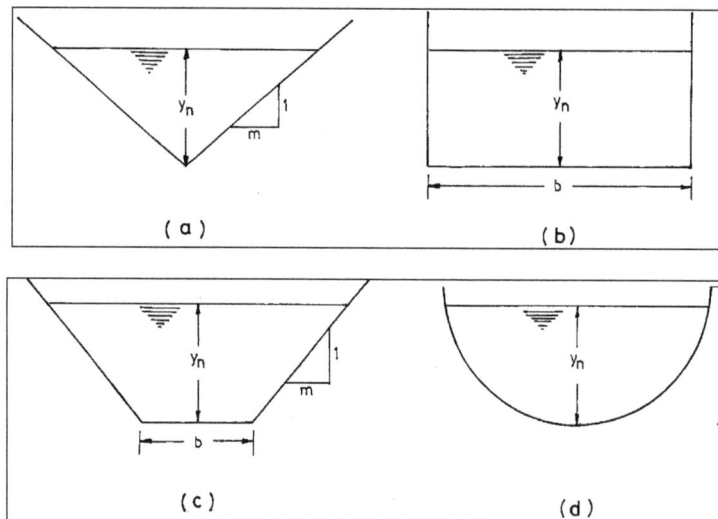

Canal sections: (a) triangular section, (b) rectangular section, (c) trapezoidal section, (d) circular section.

For a triangular section of side slope m horizontal to 1 vertical, Equations $C = c_e A + c_L P + c_r \int_0^{y_n} a \, d\eta$.

and $\phi = Q + 2.457 \, A\sqrt{gRS_0} \, \ln\left(\frac{\varepsilon}{12R} + \frac{0.221v}{R\sqrt{gRS_0}} \right)$. were reduced to,

$$\text{minimize } C_* = 2c_{L*}y_{n*}\sqrt{1+m^2} + my_{n*}^2 + \frac{c_{r*}my_{n*}^3}{3}$$

$$\text{subject to } \phi = 1.737\frac{m^{1.5}y_{n*}^{2.5}}{(1+m^2)0.25}\ln\left(\frac{\varepsilon_*(1+m^2)^{0.5}}{6my_{n*}} + \right.$$

$$\left. +0.625v_*\frac{(1+m^2)^{0.75}}{(my_{n*})^{1.5}}\right) + 1 = 0.$$

Using the optimization algorithm for a number of values of ε_*, v_*, c_{L*}, and cr_* varying in the ranges,

$$10^{-6} \le \varepsilon_* \le 10^{-3},$$

$$10^{-7} \le v_* \le 10^{-5},$$

$$0 \le c_{L*} < \infty,$$

$$0 \le c_{r*} \le or\, 0 \le c_{r*} \le 50c_{L*}$$

a large number of optimal sections were obtained. Analysis of these sections yielded the following empirical equations:

$$C^* = 1.42486c_LL + 0.25302c_eL^2 + 0.03965c_rL^3,$$

$$m^* = 1 + \frac{0.30389c_rL^2}{c_eL + 15.0491c_L},$$

$$y_n^* = 0.50301L\left(1 + \frac{0.13973c_rL^2}{c_eL + 15.03886c_L}\right)^{-1},$$

where * indicated optimality; and L = length scale given by,

$$L = \lambda(\varepsilon_* + 8v_*)^{0.04}$$

Following the above procedure for rectangular, trapezoidal, and circular canal sections as depicted in figures, generalized optimal equations for all the four canal shapes were expressed as:

$$C^* = k_{cL}c_LL + k_{ce}c_eL^2 + k_{cr}c_rL^2$$

$$m^* = k_{m0} + \frac{k_{mr}c_rL^2}{c_eL + k_{mLc_L}}$$

$$b^* = k_{b0}L + \frac{k_{br}c_rL^3}{c_eL + k_{bLc_L}}$$

$$D^* = k_{D0}L + \frac{k_{Dr}c_rL^3}{c_eL + k_{DLc_L}},$$

$$y_n^* = k_{y0}L\left(1 + \frac{k_{yr}c_rL^2}{c_eL + k_{yLc_L}}\right)^{-1},$$

where b = bed width (m); D = diameter (m); k = section shape coefficients in which the first subscripts c, m, b, D and y denote cost function, side slope, bed width, diameter and normal depth respectively; and the second subscript L, e, r, and o denote lining, earthwork, and earthwork increasing rate, and the case c_r = 0, respectively. Table lists the section shape coefficients.

For a given set of data, the use of Equations $L = \lambda \left(\varepsilon_* + 8v_* \right)^{0.04}$, along with table results in the optimal canal section. For this topic the average flow velocity V can be obtained by Equation $Q = AV$. This velocity should be greater than the nonsilting velocity but less than the limiting velocity V_L. The limiting velocity depends on the lining material as given in table. If V is greater than V_L, a superior lining material having larger V_L should be selected.

Table: Properties of optimal canal sections.

Entity	Section shape	Section shape			
	coefficients	Triangular	Rectangular	Trapezoidal	Circular
(1)	(2)	(3)	(4)	(5)	(6)
Side slope	k_{mo}	1.00000		0.57735	
	k_{mL}	15.0491		14.2772	
	k_{mr}	0.30389		0.12485	
Bed width or	k_{bo} or k_{Do}		0.71136	0.43407	0.78065
diameter	k_{bL} or k_{DL}		15.0284	14.2425	13.6232
	k_{br} or k_{Dr}		0.22772	0.15121	0.19375
Normal depth	k_{yo}	0.50301	0.35568	0.37592	0.39032
	k_{yL}	15.0389	15.0234	14.2274	12.9379
	k_{yr}	0.13973	0.30657	0.22332	0.12631
Cost	k_{ce}	0.25302	0.25302	0.24476	0.23932
	k_{cL}	1.42486	1.42396	1.30367	1.22652
	k_{cr}	0.03965	0.03961	0.03723	0.03712

Table: Limiting velocities.

Lining material	Limiting velocity (m s−1)
(1)	(2)
Boulder	1.0–1.5
Brunt clay tile	1.5–2.0
Concrete tile	2.0–2.5
Concrete	2.5–3.0

Salient Points

Equation $C^* = k_{c_L} c_L L + k_{c_e} c_e L^2 + k_{c_r} c_r L^2$ gives the optimal cost per unit length of the canal. Dividing a canal alignment into various reaches with constant Q, So, v, ε, c_e, c_L, and c_r, the canal cost F can be worked out by using Equation $C^* = k_{c_L} c_L L + k_{c_e} c_e L^2 + k_{c_r} c_r L^2$. Thus,

$$F = \sum_{i=1}^{n} \left(k_{c_L} c_{Li} L_i + k_{c_e} c_{ei} L_i^2 + k_{c_r} c_{ri} L_i^3 \right),$$

where n = number of reaches. Considering various alignments the minimum cost alignment can be finalized. Further by changing the shape of cross-section, the coefficients k_{cL}, k_{ce}, and k_{cr}, occurring in Equation $F = \sum_{i=1}^{n} \left(k_{c_L} c_{Li} L_i + k_{c_e} c_{ei} L_i^2 + k_{c_r} c_{ri} L_i^3 \right)$, can be altered. Thus, the canal shape yielding minimum F can be found out. Similarly, by trying various types of linings having different roughness, one may arrive at the appropriate lining. Thus, Equation $F = \sum_{i=1}^{n} \left(k_{c_L} c_{Li} L_i + k_{c_e} c_{ei} L_i^2 + k_{c_r} c_{ri} L_i^3 \right)$, can be used at the planning stage of a water resources project.

Equations $m^* = k_{m0} + \dfrac{k_{mr} c_r L^2}{c_e L + k_{mLc_L}} - y_n^* = k_{y0} L \left(1 + \dfrac{k_{yr} c_r L^2}{c_e L + k_{yLc_L}} \right)^{-1}$ indicate that for $c_r = 0$, the optimal section is the minimum area section. However, with the increase in c_r the canal section becomes wide and shallow. On the other hand, with the increase in ce and/or c_L, the canal section approaches to the corresponding minimum area section.

Design Steps

For the design g = 9.80 m s^{-2}; v = 1 × 10^{-6} m^2 s^{-1} (water at 20 °C); and ε = 1 mm are adopted.

Using Equation (11), λ = 24 m; using Equations (12b)–(12c) $c_L^* = 0.5$; and $c_r^* = 3.429$; and using Equation $L = \lambda \left(\varepsilon_* + 8 v_* \right)^{0.04}$ L = 16.06 m.

For a trapezoidal section previous table gave:

$k_{mo} = 0.57735$; $k_{mL} = 14.2772$; $k_{mr} = 0.12485$; $k_{bo} = 0.43407$; $k_{bL} = 14.2425$; $k_{br} = 0.15121$;

$k_{yo} = 0.37592$; $k_{yL} = 14.2274$; $k_{yr} = 0.22332$; $k_{ce} = 0.24476$; $k_{cL} = 1.30367$; and $k_{cr} = 0.03723$.

With these coefficients Equations $m^* = k_{m0} + \dfrac{k_{mr} c_r L^2}{c_e L + k_{mLc_L}}$, $b^* = k_{b0} L + \dfrac{k_{br} c_r L^3}{c_e L + k_{bLc_L}}$ and

$D^* = k_{D0} L + \dfrac{k_{br} c_r L^3}{c_e L + k_{DLc_L}}$, yielded: $m^* = 0.602$; $b^* = 7.461$ m; and $y_n^* = 5.783$ m.

Further Equation $C^* = k_{c_L} c_L L + k_{c_e} c_e L^2 + k_{c_r} c_r L^2$ gave: lining cost per meter $C_L = 251.2433 c_e$; and the excavation cost per meter $C_e = 63.1294 c_e + 22.0309 c_e = 85.1603 c_e$. Thus, the canal cost per meter $= C_e + C_L = 336.4036 c_e$. It can be seen that the lining shares the major portion of the total cost. These dimensions yield $A = y_n (b + m \, y_n) = 63.280 \, m^2$ Thus, $V = 125 / 63.280 = 1.975 \, m \, s^{-1}$ which is within the permissible limit.

Basic Components of Barrage

The only difference between a weir and a barrage is of gates, that is the flow in barrage is regulated by gates and that in weirs, by its crest height. Barrages are costlier than weirs.

Weirs and barrages are constructed mostly in plain areas. The heading up of water is affected by gates put across the river. The crest level in the barrage (top of solid obstruction) is kept at low level. During flood, gates are raised to clear of the high flood level. As a result there is less silting and provide better regulation and control than the weir.

Components of Barrage

Shutters or Gates:

Weirs are provided either with shutters or counter balanced gates to maintain pond level. A shuttered weir is relatively cheaper but locks in speed. Better control is possible in a gated weir (barrage). Their function is:

a. To maintain pond level.

b. To raise the water level during low supplies. In case of higher floods, shutters are dropped down and overflow takes place while in case of gated weir, gates are raised during floods.

Main barrage portion:

a. Main body of the barrage, normal RCC slab which supports the steel gate. In the X-Section it consists of :

b. Upstream concrete floor, to lengthen the path of seepage and to project the middle portion where the pier, gates and bridge are located.

c. A crest at the required height above the floor on which the gates rest in their closed position.

d. Upstream glacis of suitable slope and shape. This joins the crest to the downstream floor level. The hydraulic jump forms on the glacis since it is more stable than on the horizontal floor, this reduces length of concrete work on downstream side.

e. Downstream floor is built of concrete and is constructed so as to contain the hydraulic jump. Thus it takes care of turbulence which would otherwise cause erosion. It is also provided with friction blocks of suitable shape and at a distance determined through the hydraulic model experiment in order to increase friction and destroy the residual kinetic energy.

Divide Wall

It is a long wall constructed at right angle to the weir axis. It is extended up to the upstream end of the canal head regulator. In case of one canal off-taking from each bank of the river, one divide-wall is provided on front of each of the head regulators of the off takes. Similarly on the d/s side it should extend to cover the hydraulic hump and the resulting turbulence. The main functions are as follows:

1. To generate a parallel flow and thereby avoid damage to the flexible protection area of the undersluice portion.

2. To keep the cross-section, if any, away from the canal.

3. To serve as a trap for coarser bed material.

4. To serve as a side-wall of the fish ladder.

5. To separate canal head regulator from main weir.

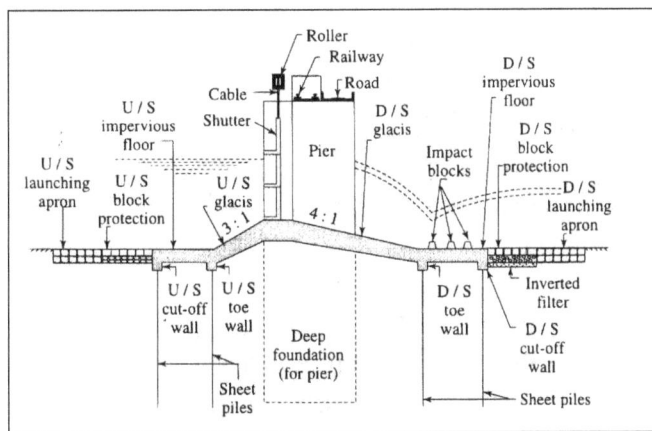

Component parts of barrage.

Fish Ladder

It is a narrow trough opening along the divide wall towards weir side provided with baffles (screen to control the flow of the liquid, sand etc.), so as to cut down the velocity of flowing water from u/s to d/s. location of fish ladder adjacent to divide wall is preferred because there is always some water in the river d/s of the under sluice only. It may be built within the divide wall. A fish ladder built along the divide wall is a device designed to allow the fish to negotiate the artificial barrier in either direction. In the fish ladder, the optimum velocity is (6-8) ft/sec.

This can be at Maralam Qadirabad & Chashma barrages. Fish move from u/s to d/s in search of relatively warm water in the beginning of water and return u/s for clear water before the onset of monsoon.

Sheet Piles

Made of mild steel, each portion being 1.5' to 2' in width and 1/2" thick and of the required length, having groove to link with other sheet piles.

There are generally three or four sheet piles. From the functional point view, in a barrage, these are classified into three types:

1. Upstream sheet piles.

2. Intermediate sheet piles.

3. Downstream sheet piles.

1. Upstream Sheet Piles:

Upstream sheet piles are located at the U/S end of the U/S concrete floor. These piles are driven into the soil beyond the maximum possible scour that may occur. Their functions are: to protect the barrage structure from scour. To reduce uplift pressure in the barrage floor. To hold the sand compacted and densified between two sheet piles in order to increase the bearing capacity when the barrage floor is defined as raft.

Functions:

- Protect barrage structure from scour.

- Reduce uplift pressure on barrage.

- To hold the sand compacted and densified between two sheet piles in order to increase the bearing capacity when barrage floor is designed as raft.

2. Intermediate sheet piles:

- Situated at the end of upstream and downstream glacis. Protection to the main structure of barrage (pier carrying the gates, road bridge and the service bridge) in the event of the upstream and downstream sheet piles collapsing due to advancing scour or undermining. They also help lengthen the seepage path and reduce uplift pressure.

- Downstream sheet piles: Placed at the end of downstream concrete floor. Their main function is to check the exit gradient. Their depth should be greater than the possible scour.

3. Down Stream Piles:

- These are placed at the end of the d/s concrete floor and their main function is to check the exit gradient. Their depth should be greater than the maximum possible scour.

Inverted Filter

An inverted filter is provided between the d/s sheet piles and the flexible protection. It typically consists of 6" sand, 9" coarse sand and 9" gravel. The filter material may vary with the size of the particles forming river bed. It is protected by placing concrete blocks of sufficient weigh and size, over it.

Slits (jhiries) are left between the blocks to allow the water to escape. The length of the filter should be (2 × downstream depth of sheet pile).

Functions:

- It checks the escape of fine soil particles in the seepage water.

- In the case of scour, it provides adequate cover for the downstream sheet piles against the steepening of exit gradient.

Flexible Apron

A flexible apron is placed d/s of the filter of the filter and consists of boulders large enough not to be washed away by the highest likely water velocity. The protection is enough as to cover the slope of scour depth i.e. (112 × depth of scour on u/s side) and (2 × scour depth on the d/s side) at a slope of 31.

Under Sluices: Scouring Sluices

Under sluice is the opening at low level in the part of barrage which is adjacent to the off takes. These openings are controlled by gates. They form the d/s end of the still ponds bounded on two sides of divide-wall and canal head regulator.

Functions:

They perform the following functions:

- To control silt entry into the canal.

- To protect d/s floor from hydraulic jump.

- To lower the highest flood level.

- To scour the silt deposits in the pockets periodically.

- To maintain a clear and well-defined river channel approaching the canal head-regulator. A number of bays at the extreme ends of the barrage adjacent to the canal regulator have a lower crest level than the rest of the bays. The main function is to draw water in low river flow conditions due to formation of a deep channel under sluice portion. This also helps to reduce the flow of silt into the canal due to drop in velocity of river water in deep channel in front of canal regulator. Accumulated silt can be washed away easily by opening the under sluice gates due to high velocity currents generated by lower crest levels or a high differential head.

- As the bed of under sluice is not lower level than rest of the weir, most of the day, whether flow unit will flow toward this pocket => easy diversion to channel through Head regulator

- Control silt entry into channel.

- Scour the silt (silt excavated and removed).

- High velocity currents due to high differential head.

- Pass the low floods without dropping.

- The shutter of the main weir, the raising of which entails good deal of labor and time.

- Capacity of under sluices.

- For sufficient scouring capacity, its discharging capacity should be at least double the canal discharge.

- Should be able to pass the dry weather flow and low flood, without dropping the weir shutter.

- Capable of discharging 10 to 15% of high flood discharge.

References

- Canal-irrigation-uses-effects: aboutcivil.org, Retrieved 8 July, 2019
- Canal-linings-types-advantages, water-resources: theconstructor.org, Retrieved 28 April, 2019
- Minliningcost-WRM: iitd.ac.in, Retrieved 19 March, 2019
- Barrages-definition, components: aboutcivil.org, Retrieved 17 February, 2019

Chapter 6

Irrigation Hydraulics and Fluid Mechanics

Irrigation hydraulics deals with the designing and maintenance of irrigation systems in an economical and efficient manner using the principles from the field of hydraulics. The topics elaborated in this chapter will help in gaining a better perspective about the branches of irrigation hydraulics such as irrigation water management and surface irrigation hydraulics.

Irrigation Hydraulics

Water pressure in an irrigation system will affect the performance of the sprinklers. If the system is designed correctly there will be enough pressure throughout the system for all the sprinklers to operate properly. Maintaining this pressure in the system will help to ensure the most uniform coverage possible. While a consistent pressure is the primary goal, it is important to achieve this at the lowest cost. With a knowledge of hydraulics, it is possible to design a system using the smallest and therefore least expensive components while conserving sufficient pressure for optimum system performance.

1. The effect of static and dynamic pressure on sprinkler operation.

2. The forces that cause pressure to increase or decrease in an irrigation system.

3. The relationships between pressure, velocity, and flow in an irrigation system.

4. How to calculate static and dynamic pressures at various points in an irrigation system.

5. How to determine dynamic pressure losses in pipe and fittings.

Water Pressure

Water pressure in irrigation systems is created in two ways:

1. By using the weight of water (such as with a water tower) to exert the force necessary to create pressure in the system or.

2. By the use of a pump (a mechanical pressurization).

In many municipal water delivery systems both of these methods may be used to create the water pressure we have at our homes and businesses. Water tanks use gravity to create pressure. These tanks are located on a mountain top, tower, or roof top. Because these storage tanks are located above the homes they serve, the weight of the water creates pressure in the pipes leading to those homes.

A "booster" pump is used to increase the pressure where the elevation of the water storage tank is not high enough above the home to provide sufficient pressure. In other areas, the water source may be a well, lake, or canal with a pump generating the pressure.

Water pressure can be measured or expressed in several ways:

- PSI; the most commonly used method in landscape irrigation, pounds of pressure exerted per square inch (PSI),

- feet of head; equivalent to the pressure at the bottom of a column of water 1 ft. high. In this case, the unit of measurement is feet of head (ft./hd).

How Pressure is Created by the Weight of Water

What water weighs at 60° F:

- 1 cubic foot (ft.³) or 1728 cubic inches (in.3) of water = 62.37 lb.

- 1 cubic inch, (in.³) of water = 0.0361 lbs.

Water creates pressure in landscape irrigation systems by the accumulated weight of the water.

In figure, we can see a container 1 ft. high and 1 ft. wide, holding 1 ft.³ of water, would create a column of water 1 ft. high over every square inch on the bottom of the container.

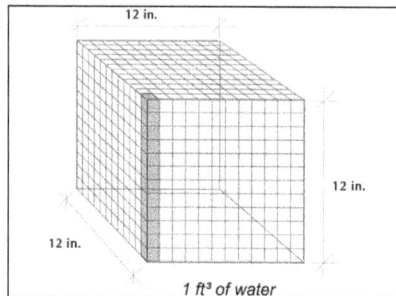

1 ft³ of water

If we look at just one of those columns, figure, we can calculate the weight of water pressing on the bottom of the column in pounds per square inch (PSI).

A column 12 in. high resting on a surface at the bottom of 1 in.2 represents a column with 12 in.3 of water.

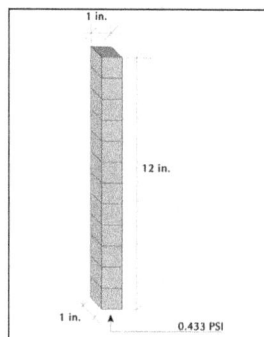

0.433 PSI

The weight of the 12 in.-high column of water is 0.433 lbs. (12 in.^{3} x 0.0361 lbs. per in.^{3} = 0.433 lbs.). Therefore, a column of water 1 ft. high will exert a pressure at the bottom of 0.433 lbs. per in.^{2} or 0.433 PSI. This is a very important number because it means that as our column of water gets higher, every 1 ft. of height added will increase the pressure at the bottom by 0.433 PSI.

For example, a column of water 2 ft. high creates a pressure at the bottom of 0.866 PSI (0.433 PSI/ ft. x 2 ft. = 0.866 PSI).

Important Facts

This gives us some important facts to remember.

Memorize these facts:

- A column of water 1 ft. high = 1 foot of head = 0.433 PSI.

- 1.0 PSI equals the pressure created by a column of water 2.31 ft. high, or 1 PSI = 2.31 ft. of head (ft./head).

- A column of water 1 ft. high creates 0.433 PSI at the bottom, or 1 ft./head = 0.433 PSI.

Effects of Pressure on the Shape and Size of the Container

The shape or size of the container does not make any difference in the pressure at the bottom, as seen in figure. Because we are measuring the weight of water in a column resting on 1 in.^{2} regardless of the container's size or shape, pressure at an equal depth will be the same no matter what the shape or size of the container.

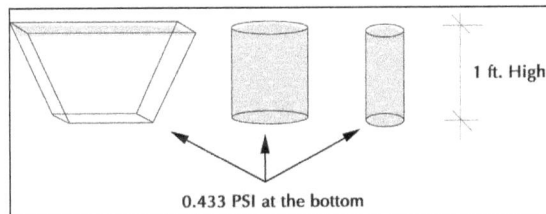

While at first this does not seem possible, let's look at two examples that will help us to better understand this concept. First, consider the example of diving into a swimming pool or lake. When you dive below the surface of a lake or pool, the deeper you dive the more pressure builds up on your ears. The amount of increased pressure on your ears does not change with the shape of the pool nor does it change depending on whether you are diving into a backyard pool or a large lake. The pressure at any depth in that pool or lake is dependent upon the height of the column of water above that point – not on the shape or size of the pool.

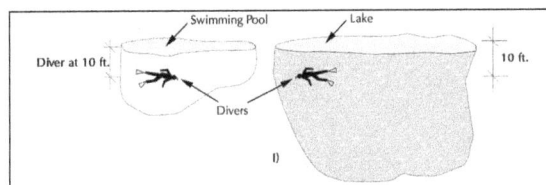

Pressure on Diver in Pool and Lake 4.33 PSI (10 ft. x 0.433 PSI/ft.) = 4.33 PSI.

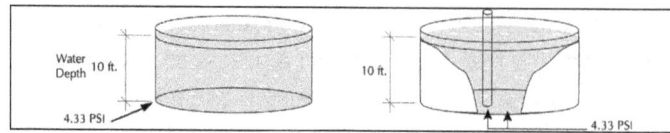

In the diagram, figure, we can determine the pressure at the bottom of the tank to be 4.33 PSI (0.433 PSI/ft. of depth x 10 ft. water depth). If we lower a 10 ft. long section of irrigation pipe (open on both ends) down into the tank, the pressure at the bottom of the pipe will be the same as that of the surrounding water, 4.33 PSI. What may still be confusing about these concepts is that intuitively we know the total weight of the water in the pipe and in the larger tank is not the same – and that is true. However, we measure pressure as the force on 1 square inch (pounds per square inch, PSI) not total weight.

Role of Pressure in Designing Irrigation System

When designing landscape irrigation systems, for every 1 ft. of elevation change there will be a corresponding change of pressure of 0.433 PSI.

Static and Dynamic Pressure

There are two classifications of water pressure.

Static and dynamic1 pressure:

- Static pressure is a measurement of water pressure when the water is at rest. In other words, the water is not moving in the system.

- Dynamic pressure is a measurement of water pressure with the water in motion (also known as working pressure).

Factors Affecting Static Pressure

Static pressure is created either by elevation change or by a pump. each foot of elevation change results in a 0.433 PSI change in pressure. As we can see in the following diagrams, the change in elevation that we are concerned with is the change in vertical elevation only, not in the length of pipe. Because water exerts pressure equally in all directions, the length of pipe will not affect the static pressure.

If we return to our tank of water, figure, we can see that inserting a fourteen foot pipe into the tank of water at an angle does not affect the pressure at the bottom of the tank or pipe. Static pressure is not affected by the length of the pipe, only by elevation change.

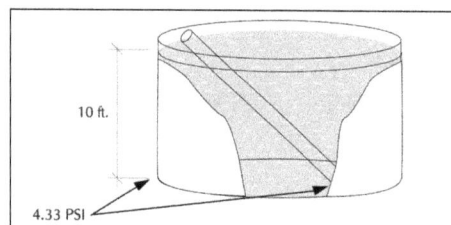

Length of Pipe and Static Pressure.

We can see the effect of elevation change on static pressure in an irrigation system in Fig In the example in Fig. , the static pressure at the water meter is 60 PSI. Since the control valve is below the water meter by 8 ft., the static pressure is increased by 3.46 PSI. (8 ft. x 0.433 PSI/ft. = 3.46 PSI).

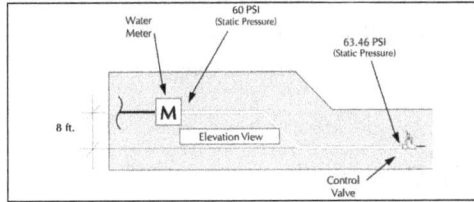

The effect of elevation loss on static pressure.

8 ft. x 0.433 PSI/ft. = 3.46 PSI

60 PSI + 3.46 PSI = 63.46 PSI static pressure

Going uphill reverses the process: for every 1 ft. of vertical elevation gain the static pressure will drop by 0.433 PSI.

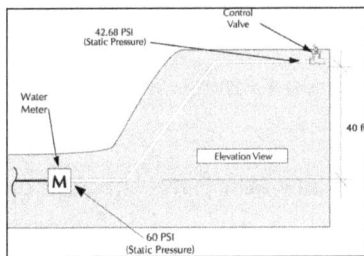

The effect of elevation gain on static pressure.

40 ft. x 0.433 PSI/ft. = 17.32 PSI

60 PSI - 17.32 PSI = 42.68 PSI static pressure

Static pressure is not affected by the size or length of pipe. Both diagrams in Fig. 8 illustrate a control valve 40 ft. above a water meter. In the first case the main line from the meter to the valve is 100 ft. of two inch pipe and in the second it is 250 ft. of one inch pipe. The static pressure at each control valve is 42.68 PSI. Only the vertical elevation change affects the static pressure.

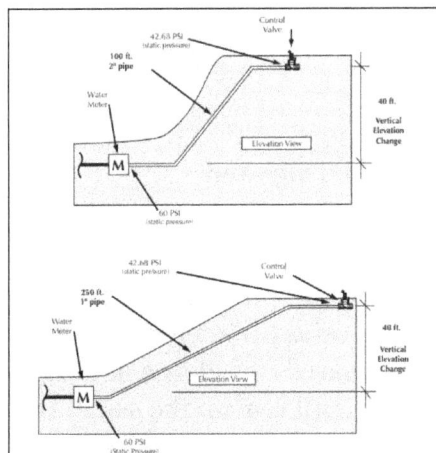

Vertical elevation change and static pressure.

Water Movement in Irrigation Systems

When water moves through an irrigation system it is said to be in a dynamic state. The movement of water is described in terms of velocity (the speed at which it is moving) and flow (the amount of water moving through the system). The velocity is measured in feet per second (FPS) and the flow is measured in gallons per minute (GPM). Dynamic water pressure is measured in the same units as static pressure (PSI).

Factors Affecting Dynamic Pressure

So far we have discussed the factors that affect changes in static pressure. This topic explain factors that affect dynamic pressure.

Dynamic pressure is affected by the following factors:

1. Change in elevation (change in elevation affects static and dynamic pressure in the same way) .

2. Friction losses in pipe, valves and fittings (pressure loss is caused by water moving through the system).

3. Velocity head (the force required to make water move within the system; this is a minor loss).

4. Entrance losses (pressure lost as water flows through openings; this is also a minor loss).

Friction Loss in Pipe

When measuring dynamic pressure at any point in a landscape irrigation system, we must first determine the static pressure at that point and then subtract the pressure losses due to the movement of water.

As water moves through an irrigation system, pressure is lost because of turbulence created by the moving water. This turbulence can be created in pipes, valves, or fittings. These pressure losses are referred to as "friction losses."

There are four factors that affect friction losses in pipe:

1. The velocity of the water,

2. The inside diameter of the pipe,

3. The roughness of the inside of the pipe and

4. The length of the pipe.

1. Velocity is the speed at which water moves through the system and it is measured in feet per second (FPS). Water moving in the pipe causes turbulence and results in a loss of dynamic pressure. Increasing the velocity will cause increased turbulence and increased pressure losses. In figure, the inside diameter, roughness, and length remain the same. However, due to increased velocity (FPS), there is a greater dynamic pressure loss.

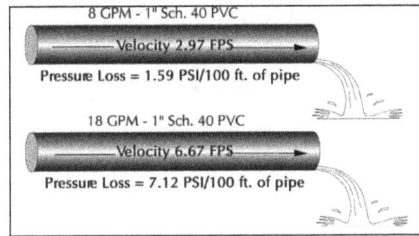

Effect of velocity on dynamic pressure.

When velocity increases, pressure loss increases. When the velocity is increased from 2.97 FPS to 6.67 FPS, the pressure lost in 100 ft. of pipe increases from 1.59 PSI to 7.12 PSI. The velocity typically increases when 1) the flow is increased, such as when additional sprinklers are added to an existing line or 2) a smaller pipe is used with the same flow (GPM).

2. Inside Diameter (I.D.) of the pipe: a smaller inside pipe diameter proportionally increases the amount of water in contact with the pipe surface. This increased contact increases the turbulence and consequently increases the dynamic pressure loss. In figure below, the velocity, length, and roughness remain the same but the inside pipe diameter is reduced. The reduced i.d. results in increased turbulence and reduced dynamic pressure.

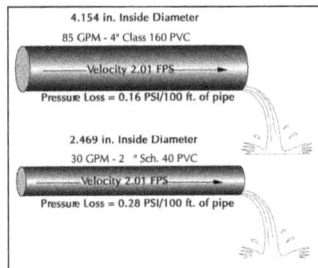

Effect of inside diameter on dynamic pressure loss.

Even with a smaller flow and the same velocity more turbulence was created in the small pipe because there was a greater percentage of the water in contact with the surface.

3. Roughness of the inside wall of the pipe is the third factor that affects friction loss in pipe. Pipe wall roughness is rated by a "C" factor. The lower the value of C, the rougher the inside wall of the pipe (in standard steel pipe C = 100; in PVC pipe C = 150.) The rougher the inside, the more turbulence created and the greater the pressure loss.

In figure below the velocity, volume and inside diameter remain the same. As the roughness of the inside of the pipe increases (standard steel has a rougher pipe wall than PVC), there is an increase in turbulence, resulting in a greater pressure loss.

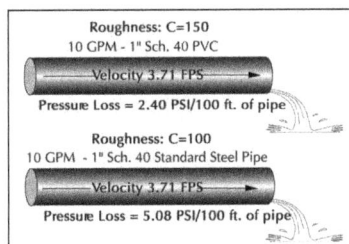

Effect of pipe wall roughness on dynamic pressure loss.

4. LENGTH is the fourth factor affecting friction losses in pipe. The greater the distance, the greater the cumulative effect of the first three factors (velocity, inside diameter, and roughness). In figure we see the direct relationship between increased length and increased pressure loss. The total pressure loss doubles as the length of the pipe doubles.

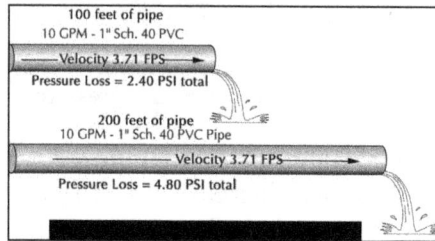

Effect of pipe length on dynamic pressure loss.

These four factors affecting pressure loss in pipe were used to develop formulas2 for calculating the pressure loss associated with various types of pipe. Several formulas were developed; the most common in landscape irrigation hydraulics is the Hazen-Williams formula. The Hazen-Williams formula can be represented as:

$$H_f = 0.090194 \left(\frac{100}{c} \right)^{1.852} \frac{Q^{1.852}}{d^{4.866}}$$

Where Hf = pressure loss in pounds per square inch (PSI) per 100 ft. of pipe

    C = roughness factor

    Q = flow in gallons per minute (GPM)

    d = inside pipe diameter in inches

Since these formulas are somewhat cumbersome, we will rely on charts developed using the Hazen-Williams formula.

Use of Pressure Loss Charts

Figure below represents a portion of one of the pressure loss.

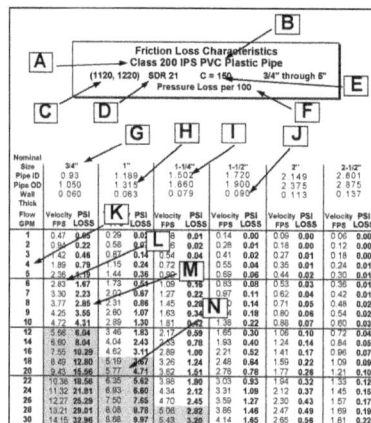

Explanation of a pressure loss chart.

The components of the typical Friction Loss Chart are described below:

a. Type of pipe represented in the chart.

b. IPS – Iron Pipe Size - indicates that the pipe's outside diameter dimensions correspond to that of iron pipe. All IPS PVC pipe of the same nominal size will have the same outside diameter. For example: all ½" PVC irrigation pipe will have an outside diameter of 0.840 in.; thus all ½" slip fittings will fit on the outside of all types of ½" PVC pipe.

c. (1120, 1220) – Represents a designation for the specifications of the plastic pipe.

d. SDR – Standard Dimension Ratio – indicates the pipe's wall thickness as a ratio of the outside diameter. Outside diameter of 1" pipe is 1.315 in. If you divide 1.315 by the SDR, 21, it will give you a minimum wall thickness. Minimum wall thickness for 1" Class 200 PVC pipe 1.315/21 = 0.063 in. Class-rated pipes (SDR pipes) maintain a uniform maximum operating pressure across all pipe sizes. This is not true of schedule rated pipes such as Schedule 40 PVC. In schedule rated pipes the maximum operating pressure decreases as pipe size increases.

e. C=150 – indicates the value of the C factor, which is a measure of the roughness of the inside of the pipe. The lower the number, the rougher the inside of the pipe and the greater the pressure loss. For PVC, C = 150; Galvanized Pipe C = 100.

f. Designated pressure losses shown in the chart are per 100 ft. of pipe.

g. Size – indicates the "nominal" pipe size. Nominal means "in name only," and none of the actual pipe dimensions are exactly that size. For example, in the ¾" pipe, none of the dimensions are actually ¾"

h. OD – outside pipe diameter in inches.

i. ID – inside pipe diameter in inches.

j. Wall Thick – wall thickness in inches. K) Flow (GPM) – flow rate in gallons per minute.

k. Velocity (FPS) – speed of water in feet per second at the corresponding flow rate.

l. PSI Loss – pressure loss per 100 ft. of pipe in pounds per square inch at the corresponding flow rate.

m. The shaded area on the chart designates those flow rates that exceed 5 FPS. It is recommended that caution be used with flow rates above 5 FPS in main lines where water hammer will be a concern.

Uses of Charts in Irrigation Hydraulics

These charts are used to:

a. Determine the pressure loss in pipe due to friction losses.

b. Determine the velocity at various flow rates.

c. Use pressure losses and/or velocities to determine appropriate pipe sizes.

Determining Dynamic Pressure Losses in Pipe

When calculating dynamic pressures we use the following factors:

1. Pressure change due to elevation change.

2. Pressure loss due to friction losses in the pipe.

3. Pressure losses in valves, meters, etc. (These losses are determined by the manufacturer and listed in product literature or technical charts.)

4. Pressure losses due to fittings.

The following example illustrates how the dynamic pressure at a given point in a landscape irrigation system is determined. Pressure change due to the change in elevation is calculated and the friction losses are subtracted from the sub-total.

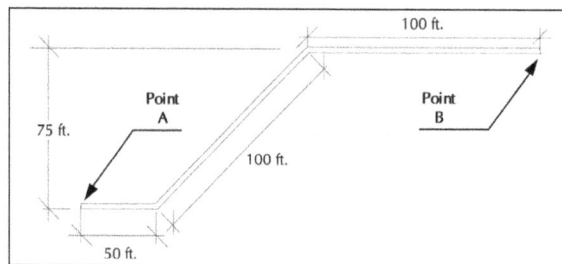

Comparing dynamic pressure at two points in an irrigation system.

In figure, all the pipe is 1¼" Class 200 PVC and the flow is 18 GPM from point A to point B. At point A a pressure gauge reading indicates 85 PSI. In order to determine the dynamic pressure at point B, first find the pressure change due to change in elevation and then combine that with the friction loss in the pipe.

Pressure loss due to the higher elevation at point B (an elevation gain):

>   75 ft. x 0.433 PSI per ft. of elevation change
>
>   = 32.48 PSI less at point B

PSI loss in pipe:

>   (50 ft. + 100 ft. + 100 ft.) x (1.24 PSI loss per 100 ft. ÷ 100 ft.) = PSI loss
>
>   PSI loss in pipe = 250 ft. x 0.0124 PSI loss per ft.
>
>   PSI loss in pipe = 3.10 PSI [total due to friction loss]
>
>   85.00 PSI pressure at point A
>
>   -32.48 PSI due to elevation change
>
>   52.52 PSI subtotal at point B
>
>   -3.10 PSI due to friction loss in pipe from point A to point B
>
>   49.42 PSI dynamic pressure at point B

General Principles of Water Flow in an Irrigation System

Water in an irrigation system has energy called pressure. The pressure may be created by the weight of a column of water; static pressure, or the pressure may be created by a pump. In the following section, you will see that as water flows through an irrigation system all of the energy or pressure that is available at the source is expended. Some of the pressure is expended as friction losses in pipes, valves, and fittings and some is used to create velocity and flow. The flow and velocity will increase until all the pressure available at the source is consumed as friction losses or used to create velocity. For example, an irrigation system with 50 PSI at the source will expend all 50 PSI between the source and the point where the water has left the system (an open pipe or sprinkler nozzle). The quantity (GPM) and velocity (FPS) will increase until the cumulative pressure losses, from the source to the outlet, equal the pressure available at the source.

In other words, every pound of increased pressure at the water source will cause more water to flow through the outlet. As the flow increases so does the velocity resulting in increased pressure losses. These increased pressure losses will equal the increase in pressure at the source.

Factors Affecting Flow in an Irrigation System

The flow in an irrigation system is determined by three factors:

1. The pressure (PSI) available at the source,

2. The pressure losses from the source to the outlet(s), and

3. The size and number of outlets.

In our examples we start with a tank filled with water. It is 115.5 ft. from the water surface to the tank outlet. The water pressure created at the tank outlet is 50 PSI (0.433 x 115.5 = 50).

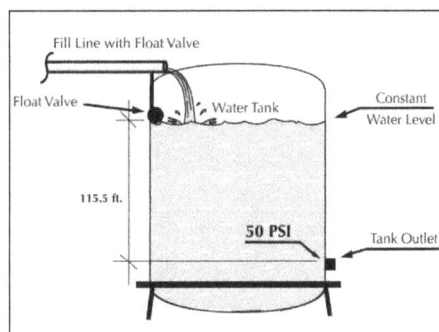

Water pressure at water tank outlet.

Effects of Pressure Available at the Source on Flow

The 50 PSI at the tank outlet, shown in figure, is created by a water tank full to 115.5 ft. above the tank outlet (115.5 ft. x 0.433 PSI per ft. = 50 PSI). In the first example, we have opened a fill-line and used a float valve to maintain a constant water level allowing water from the fill-line to replace water leaving through the tank outlet.

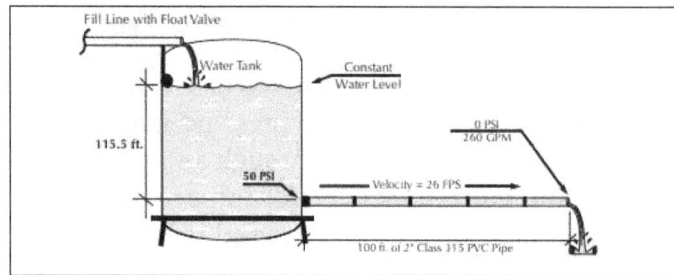

The effect of adding pipe to the system

With the 2" Class 315 PVC discharge line wide open (unrestricted flow), the water velocity (FPS) increases until all of the available pressure is lost to friction in the pipe, fittings, or other losses. The increased velocity creates increased pressure losses until the pressure losses equal the pressure available at the source. Under these conditions the velocity is 26 FPS and the flow is 260 GPM. This is greater than the velocities and pressure losses listed in friction loss charts. Mathematical formulae were used to calculate the velocity and flow.

Now where did that pressure go? One of the most difficult concepts to understand is how water that has left a pipe, like the one in the diagram above, no longer has any pressure. Remember, the water velocity in the pipe will increase until all the pressure available at the source is consumed. Consider the water leaving the "fill line" at the top of the tank. As this water leaves the pipe and flows into the top of the tank, it is no longer under pressure.

Another example is water from a hose that flows out onto the ground. The water velocity in the hose increases until all the pressure available at the source has been consumed. When the water flows out onto the ground, it is no longer under pressure.

No one should design an irrigation system with the velocity and flow this high; therefore, 26 FPS and 260 GPM are not even listed on the normal pressure loss/velocity charts and were calculated by formula instead. As we will see later, the flow and velocity are kept to much lower levels by regulating the number of sprinklers on a pipe or valve, which, in turn, regulates the flow.

Next we look at what happens when the water pressure at the source drops. In figure, the fill-line has been shut off and the water level in the tank allowed to drop. When the level in the tank has dropped to a depth of 46.19 ft. above the tank outlet the pressure at the outlet is 20 PSI (46.19 x 0.433 = 20 PSI). As might be imagined, the flow through the pipe outlet has been reduced. With the pressure at 20 PSI, the discharge is approximately 160 GPM at a velocity of 16 FPS. The drop in flow resulted from the drop in pressure at the source.

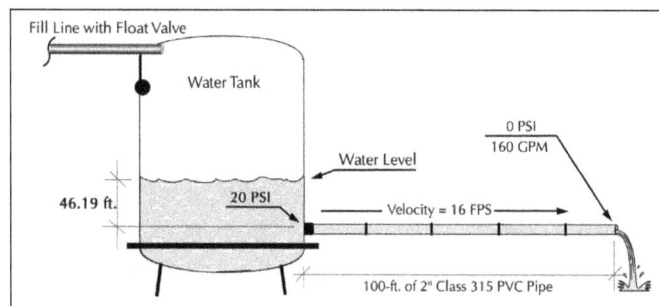

The effect of reducing the water level in the tank.

A pressurized irrigation system will dissipate all the pressure at the source by the time the water leaves the discharge point. As shown in figure, the water velocity and flow increase in the pipe until the total pressure losses incurred, from the tank to the pipe outlet, equal the pressure available at the source.

Pressure at the source directly affects the velocity and flow in an irrigation system. At 20 PSI at the source, the velocity is 16 FPS and the flow is 160 GPM. If the tank is filled back to 115.5 ft., the flow increases until there is a total of 50 PSI in pressure losses from the source to the discharge (260 GPM at 26 FPS).

How Pressure Losses from the Source to the Outlet Affect Flow

In figure, there is a constant pressure at the source of 50 PSI and an unrestricted flow through 100 ft. of 2" Class 315 PVC pipe. Under these conditions, the amount of water that will flow through the pipe is approximately 260 GPM at 26 FPS. With 50 PSI at the source, the flow of water through the pipe increases until it reaches 26 FPS. At 26 FPS.

With 50 PSI at the source, the flow of water through the pipe increases until it reaches 26 FPS. At 26 FPS pressure losses from the inlet of the pipe to the outlet equal the pressure available at the source. Figure and the table below show how and where the 50 PSI at the source is lost.

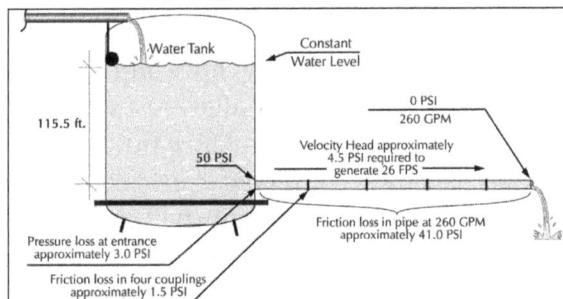

Pressure losses from source to outlet.

Friction loss in pipe – At 26 FPS through a 2" Class 315 PVC pipe (this is above the velocities and pressure losses listed in the pressure loss charts and was computed using the mathematical formulæ for pressure losses in pipe): Approximately 41.0 PSI loss.

Friction loss in fittings – Pressure lost in the four couplings (PVC pipe is made in 20-ft. sections requiring four couplings to assemble 100 ft. of pipe): Approximately 1.5 PSI loss.

Velocity head – This is the pressure required to generate 26 FPS (amount of pressure required to move the water through the pipe at 26 FPS): Approximately 4.5 PSI loss.

Entrance losses – Pressure lost as the water enters the pipe: Approximately 3.0 PSI loss.

Total pressure lost or consumed: 50 PSI.

If there is an increase in pressure (PSI) at the source, the rate of flow (GPM) and velocity (FPS) will also increase. When the flow and velocity are increased, the pressure losses from friction, velocity head and entrance losses also increase. The rate of flow increases until all the additional pressure is used to create a higher flow (GPM) and velocity (FPS). In our examples, the flow at 20 PSI is approximately 160 GPM. If the pressure is increased by 30 PSI, from 20 to 50 PSI at the source, the flow will increase by 100 GPM from approximately 160 GPM to 260 GPM. The increase of 100 GPM in the flow will increase pressure losses by 30 PSI, so that by the time the water has left the pipe, all the pressure available at the source will have been used.

Some Practical Examples:

1. The laminar flow drip emitter: If the pipe is small enough and long enough, the pressure loss will be so great that the water will just drip out. This is how some drip emitters work. Long, small water pathways (like pipes) inside the emitters cause so much pressure loss that very little velocity or flow remain at the discharge point.

2. Nozzle size controls flow: In a sprinkler system, the water flow is less than that from an open-ended pipe. Flow is controlled by limiting the number of sprinklers per control valve and the size of the sprinkler nozzles. The nozzles are smaller than the open pipe. The smaller nozzles control the flow of water in the system. Because of the reduced flow and velocity there is reduced pressure loss from the water source to the sprinkler head. The pressure available at the sprinkler is expended as the water escapes through the small nozzle. At the nozzle, the water velocity increases as it exits at the nozzle, and this increased velocity dissipates the remaining pressure. This increased velocity throws the water up to a hundred feet or more, depending on the sprinkler design.

3. The "thumb on the hose" trick: Turn on a long hose and the water barely comes out. If you put your thumb over the end of the hose and allow just a tiny amount of water to flow past your thumb, you have reduced the velocity and flow through the hose, which results in less pressure loss. The pressure you have conserved is now converted to a higher velocity as the water flows past your thumb. The higher velocity will cause the water to be thrown farther than before. This sometimes leaves a person with the impression that a smaller pipe increases pressure. In reality, the reduced flow results in more pressure remaining at the end of the hose, which in turn creates more velocity as the water leaves the hose.

The Relationship between Pressure and Flow

Figure charts the relationship between pressure at the source and flow. This relationship is shown for three sizes of PVC pipe. There are three points that should be noted about this chart.

1. Increasing pressure at the source increases the flow. On the chart the two bold dashed lines indicate the increase in flow from 160 GPM to 260 GPM as the pressure at the source is increased from 20 PSI to 50 PSI.

2. Using smaller pipe does not increase the flow. Smaller pipe sizes have less flow at any given pressure. Since decreasing pipe size does not increase the pressure at the source, the result of decreasing pipe size is a reduced flow (GPM).

3. Using smaller pipe does not result in higher pressure. Smaller pipe leads to greater pressure loss. For example, on the chart, a flow of 20 GPM in our 1" pipe would require a pressure at the source of about 9 PSI. In order to maintain the same 20 GPM flow in the smaller ½" pipe, we would need over 50 PSI at the source. Smaller pipe results in greater pressure loss, not higher pressure.

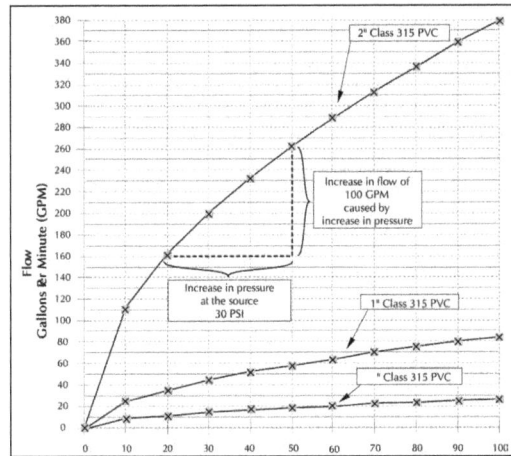
The relationship of pressure and flow.

Effects of Flow on the Size of the Outlet

At first it would seem that any 100-ft.-long section of 2" class 315 PVC pipe with 50 PSI at the source should have a flow of 260 GPM. However, it must be remembered that the size of the outlet at the discharge end of the pipe will also affect the flow of water.

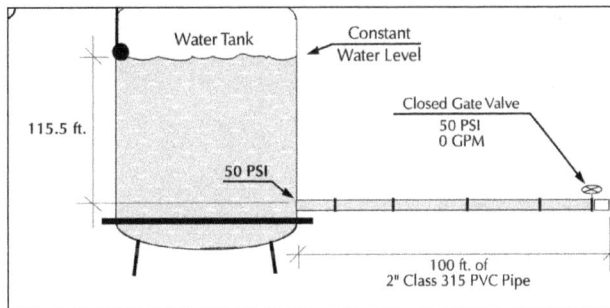
Flow is affected by a closed gate valve.

Figure is the most extreme example of this principle; the pipe has a closed gate valve that completely stops the flow. Even though the pipe is still the same size and has the same pressure at the source, there is NO flow.

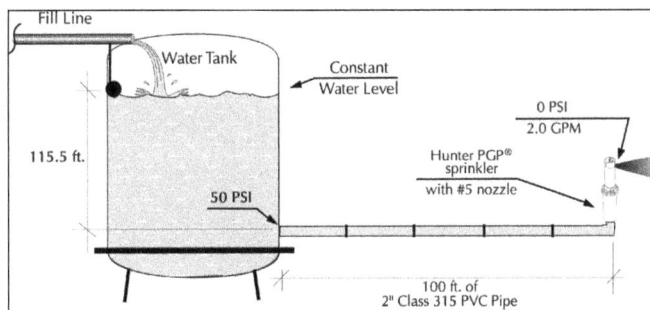
Flow is affected by a PGP sprinkler.

In figure, we have replaced the gate valve with a Hunter PGP sprinkler with a #5 Nozzle. Because the outlet is so much smaller than the pipe,with 50 PSI at the source, it will allow a flow of only 2.0

GPM5.

As shown in the previous examples, the flow in an irrigation system is controlled by:

1. The pressure (PSI) available at the source. Increasing the pressure at the source increases the flow in the system; more pressure is available to compensate for the increased pressure losses in the system.

2. The pressure losses from the source to the outlet(s). Reducing pipe size or any other factor which causes a greater pressure loss will result in reduced flow.

3. The size or number of the outlet(s). Changing the size or number of outlets, such as changing the size of the sprinkler nozzles or the number of sprinklers on a line, will change the amount of flow.

What Happens When the Water Reaches a Sprinkler Head

It is important to see how hydraulic principles apply to a landscape irrigation system. To better understand how a typical system works, let's look at a sprinkler system that has four sprinklers using 10 GPM each for a total of 40 GPM. The system has 30 PSI at the control valve and has been installed.

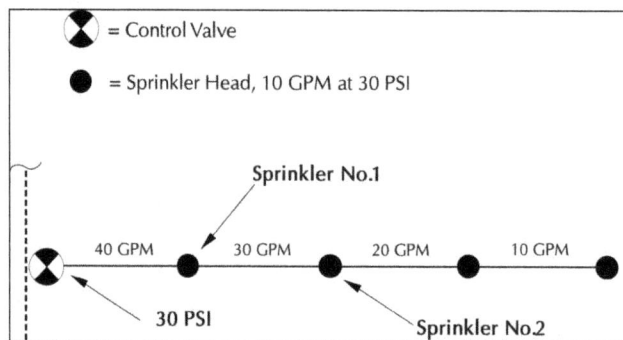

Hydraulics principles applied to an irrigation system.

When the system is in operation, 10 GPM flows from each sprinkler head. The pressure at each sprinkler remains relatively constant because the pipe and fittings have been sized to minimize pressure loss. For the sake of this example we will assume a negligible pressure loss so that each sprinkler is operating at 30 PSI. Later we will examine the pressure losses between each head.

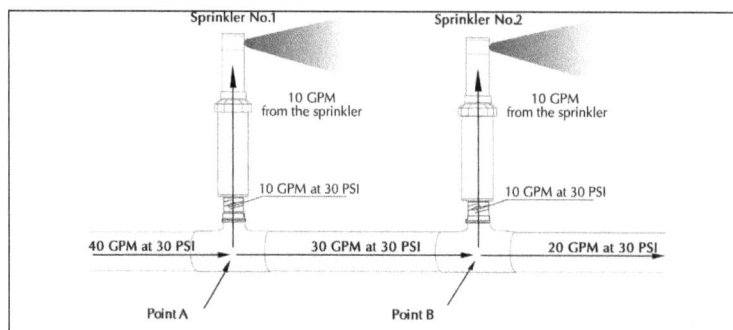

Pressure at sprinkler heads when system is operating.

At point A (below sprinkler No. 1) in figure, the flow of water splits, with 10 GPM flowing up and out of sprinkler No. 1 and 30 GPM flowing on to sprinkler No. 2. The pressure, however, does not split. Instead, the pressure is equal in both directions at point A. At point B the flow of water splits in the same way it did at point A with 10 GPM flowing to the sprinkler and 20 GPM flowing toward sprinklers farther down the line. Once again, the pressure does not split. The pressure at point B (below sprinkler No. 2) is 30 PSI both at the base of the sprinkler as well as in the direction of the sprinklers downstream.

Because this is a difficult concept to understand, figure shows a variation on the same principle.

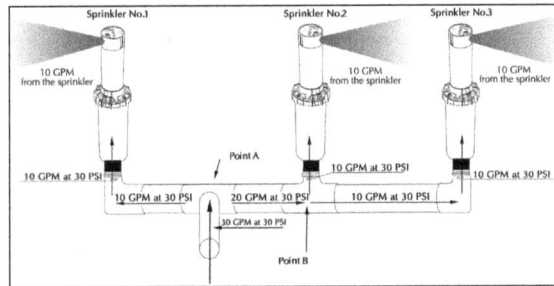

Pressure at sprinkler heads in different system layout.

In this example, 30 GPM enters the horizontal pipe at point A at 30 PSI. Ten gallons per minute then flow to sprinkler No. 1 and 20 GPM flow toward sprinklers No. 2 and No. 3. The pressure, however, at point A is pushing the water equally in both directions with 30 PSI. The volume splits but the pressure pushes equally in both directions.

The same situation occurs at point B where the water pressure is 30 PSI pushing toward sprinkler No. 2 and 30 PSI pushing toward sprinkler No. 3. In this case the 20 GPM splits, with 10 GPM going to sprinkler No. 2 and 10 GPM going on to sprinkler No. 3. The flow of water splits at each of these points but the pressure is pushing equally in each direction.

Flow and Pressure Loss in a Typical Sprinkler System

We assumed each sprinkler was operating at exactly the same pressure. In reality there are some pressure losses in the pipe between each head. In this, we will see how to calculate those pressure losses that occur between sprinklers to determine if our system will operate properly.

Figure illustrates a system with four sprinklers controlled by one valve. While there is some loss of pressure between the sprinklers, we will still assume they are delivering 6.0 GPM.

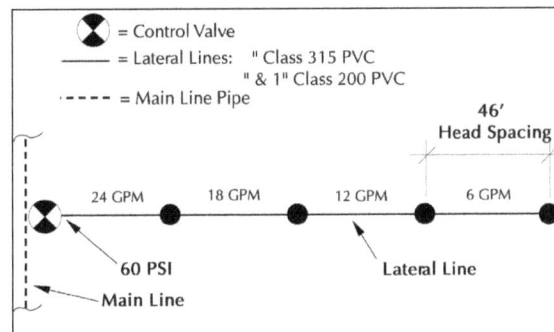

Flow and pressure loss in an irrigation system.

In order to keep pressure losses to a minimum and thereby maintain a relatively uniform pressure throughout the system, the lateral line pipe has been sized as shown in figure. (Pipe sizing is covered in another education module, but for clarification, pipe is downsized when volume is reduced because smaller pipe and fittings are less expensive.)

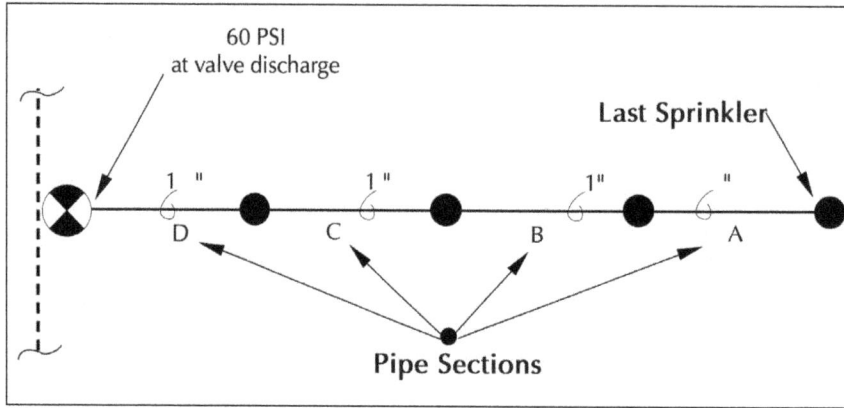

Pipe size downsized as volume is reduced.

As water flows through our sprinkler system, the pressure losses due to friction and the other factors reduce the pressure available at the sprinklers. Each sprinkler on the line has less pressure than the one before it (assuming the system is not running down a slope). It is important to maintain relatively uniform pressures for all sprinklers on a given control valve because sprinkler performance (GPM and radius of throw) will vary as the pressure at each sprinkler varies. Our design goal is to have all sprinklers controlled by one valve within +/- 10% of the pressure at which they were designed to operate.

In our sample problem, according to the sprinkler manufacturer's specifications, the sprinklers have been designed to operate at 60 PSI. This means that +/- 10% of the designed operating pressure is an operating range of 54 – 66 PSI. Since we have 60 PSI at the discharge of the valve, we must check to be sure that we have at least 54 PSI (a 10% variation) at the last head (Note: If the pipe were running down a slope, you would want to be sure you did not have more than 66 PSI). We can check to see if we have sufficient pressure at the last sprinkler by determining the pressure loss through each section of pipe. The following table, figure, will aid in these calculations.

Pipe Section	Type	Size	GPM	Length	PSI loss/ 100 ft.	Actual PSI Loss
A	Cl. 200	¾"	6	46 ft.	1.67	0.77
B	Cl. 200	1"	12	46 ft.	1.83	0.84
C	Cl. 200	1¼"	18	46 ft.	1.24	0.57
D	Cl. 200	1¼"	24	46 ft.	2.12	0.98
Total PSI loss in pipe from the valve to the last head						3.16 PSI
Pressure at the valve discharge						60.00 PSI
Pressure loss in pipe from valve to last head						-3.16 PSI
Estimated pressure loss in fittings (10% of pipe loss)						-0.32 PSI
Pressure remaining at the last head						56.52 PSI

The preceding example illustrates the pressure losses in pipe. By performing this calculation we can check to see if we have designed our system to maintain a sprinkler pressure at each head within 10% of the pressure at which the head was designed to operate. This is important because it assures us the system will operate properly before it is installed. This type of calculation would not have to be performed for every valve on a project, just for the valve most likely to have the lowest pressure (the "worst-case" scenario).

Irrigation Water Measurement

Irrigation water management begins with knowing how much water is available for irrigation.

Methods of measuring irrigation water can be grouped into three basic categories—direct, velocity-area, and constricted flow. Choice of method to use will be determined by the volume of water to be measured, the degree of accuracy desired, whether the installation is permanent or temporary, and the financial investment required.

Direct Measurement Methods

Measuring the period of time required to fill a container of a known volume can be used to measure small rates of flow such as from individual siphon tubes, sprinkler nozzles, or from individual outlets in gated pipe. Ordinarily one gallon or five gallon containers will be adequate. Small wells can be measured by using a 55 gallon barrel as the container. It is recommended that the measurement be repeated at least three and preferably five times to arrive at a reliable rate of flow per unit of time.

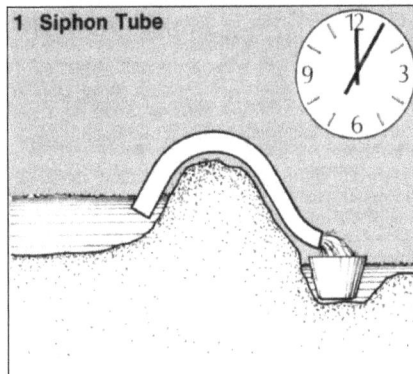

Velocity-Area Methods

Commercial flow meters are available for measuring the total volume of water flowing through a pipe. These flow meters are relatively expensive; however, they have a good degree of accuracy if properly installed and maintained. Some meters can be purchased, which will indicate instantaneous rate of flow.

The float method can be used to obtain an approximate measure of the rate of flow occurring in an open ditch. It is especially useful where more expensive installations are not justified or high degree accuracy is not required.

Select a straight section of ditch from 50 to 100 feet long with fairly uniform cross-sections. Make several measurements of the width and depth of the test cross-section so as to arrive at an average cross-sectional area. Using a tape, measure the length of the test section of the ditch. Place a small floating object in the ditch a few feet above the starting point of the test section and time the number of seconds for this object to travel the length of the test section. This time measurement should be made several times to arrive at a reliable average value. By dividing the length of the test section (feet) by the average time required (seconds), one can estimate velocity in feet per second. Since the velocity of water at the surface is greater than the average velocity of the stream, multiply the estimated surface velocity by a correction factor (0.80 for smooth lined ditches, and 0.60 for rough ditches) to obtain the average stream velocity.

To obtain the rate of flow, multiply the average crosssectional area of the ditch (square feet) times the average stream velocity (feet per second) and the answer is the rate of flow in cubic feet per second.

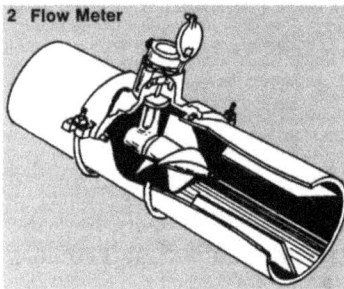

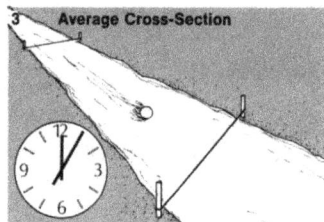

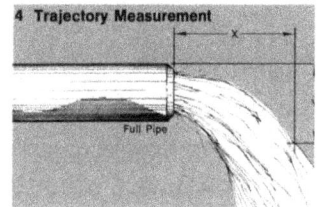

The trajectory method of water measurement is a form of velocity area calculations that can be used for determining the rate of flow discharging from a horizontal pipe flowing full. Two measurements of the discharging jet are required to calculate the rate of flow of the water. The first measurement is the horizontal distance, "X", (parallel to the centerline of the pipe) required for the jet to drop a vertical distance "Y" which is the second measurement.

By using"Y" equal to either 6 or 12 inches, the rate of flow for full pipes can be calculated by multiplying the horizontal distance "X" (in inches) times the appropriate factor for the nominal pipe diameter. The following table contains water discharge factor where "Y" is measured from the outside of the pipe as indicated in the sketch above.

Nominal Pipe Diameter	Factor When Y=6	Factor When Y=12
2"	5.02	3.52
3"	11.13	7.77
4"	17.18	13.4
6"	43.7	30.6
8"	76.0	52.9
10"	120.0	83.5
12"	173.0	120.0

EXAMPLE: A farmer has a well discharging a full 8" pipe. The horizontal distance (X) is 19" while the jet surface drops 12". What is the well yield?

Step 1: Enter the water discharge factor table at 8" nominal pipe diameter. Moving to the right and under the column headed Y = 12" we find the factor to be 52.9.

Step 2: Multiplying this factor 52.9, times the horizontal distance, 19" calculate the well yield to be 1,005 gpm.

Constriction Flow Methods

Methods employing a constriction of pre-determined dimensions are frequently used for measuring flow in irrigation canals and ditches. Constricting type measuring devices can generally be placed in one of three categories—weirs, flumes, and orifices.

Generally, only one or two measurements are required where the dimensions of the constriction are known. Using these measurements, rate of flow is determined from either a table, a graph, or by calculation. Due to the wide variety of types and sizes of constricting devices, flow tables are not included in this publication. The local County Extension Director or local Soil Conservation Service District office can obtain such tables or graphs.

Basically, a weir measures flow by causing the water to flow over a notch of pre-determined shape and dimensions. They are quite accurate when properly constructed, installed, and maintained. Weirs do have some limitations. First, they require considerable drop (difference in head) between the upstream and downstream water surfaces which is often either not available in flat grade ditches or is undesirable. Second, it is frequently necessary to construct a pool or stilling area above the weir so the water loses its velocity. Unless the water appears practically still, discharge readings will be inaccurate. Weir installations in earthen ditches can be particularly troublesome. The stilling area in the ditch above the weir frequently tends to "silt in" while excessive erosion may occur immediately downstream from the weir.

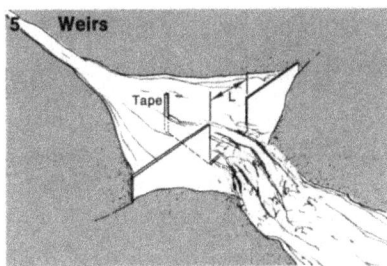

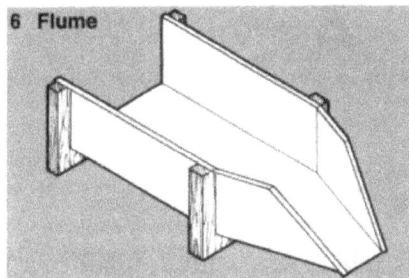

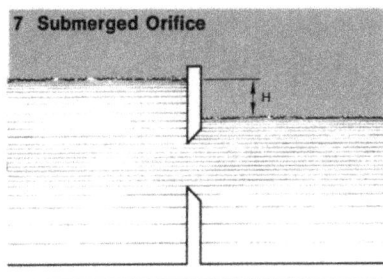

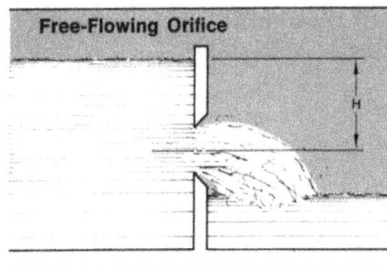

5 Weirs

6 Flume

7 Submerged Orifice

Free-Flowing Orifice

A flume measures flow by causing the water to flow through a channel of pre-determined dimensions. Flumes usually can operate with less difference in elevation between upstream and downstream water surfaces than can weirs. Like weirs, when properly installed and maintained, flumes are quite accurate means of measuring water flow.

An orifice measures water flowing through an opening of pre-determined shape and size. For a given amount of head (pressure) a specific quantity of water will flow through the opening. Orifices can be classified as "free flowing" where the flow from the orifice discharges entirely into air or "fully submerged" where the downstream water surface is above the top of the orifice and the flow discharges into water. Avoid orifices that do not flow free or are not completely submerged.

Orifice plates properly installed on open pump discharges can provide a relatively inexpensive and reasonably accurate means of measuring well discharge. It is very important that the opening in the orifice plate be accurately machined to dimension. Slight variation from specified dimensions can cause wide variation from calculated rate of flow.

The equation for calculating flow through an orifice is: $Q = K\sqrt{H}$

Where,

> Q = flow in gallons per minute.

> K = a constant dependent upon a combination of pipe size, orifice size and orifice shape, and discharge conditions.

> H = Head in inches.

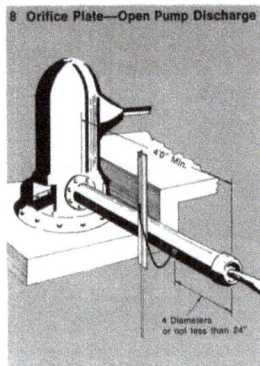

8 Orifice Plate—Open Pump Discharge

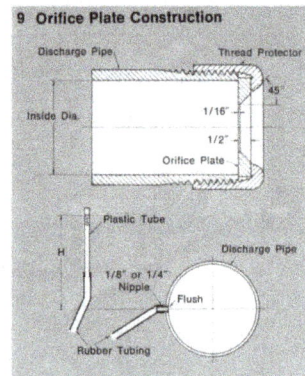

9 Orifice Plate Construction

The following table gives values of K for various combinations of orifice sizes and pipe sizes discharging into air. These values of K should be used only for orifice plates machined to the dimensions shown in Illustration.

Value of K						
Pipe Size Inches	Orifice Size in Inches					
	3	4	5	6	7	8
4	42.3					
6	33.3	63.3	1 23.0			
8		59.0	97.3	1 55.0		
10				141.0	208.0	311.0

The following graph can be used for determining the "square root" of the head (H) in inches. Having measured the head H using the glass tube and a scale, enter the graph at the left side. Move horizontally to intercept the curve and move downward to determine the square root of H ($\sqrt{H}$).

The following is an example of how to calculate flow using an orifice plate on a pump discharge. A farmer has a 6" orifice plate installed on an 8" pump discharge. The orifice plate is machined to the dimension and shape shown in the sketch. He or she determines the head "H" to be 27". Consulting the table, he or shedetermines the constant "K" for a 6" orifice in an 8" pipe to be 155. Using the graph he or she estimates the square root of 27 to be about 5.2. Substituting the values in the equation $Q = K\sqrt{H}$, Q is calculated to be 806.0 (5.2 x 155).

The final selection of the type of water measurement device will depend on the volume of water to be measured, the degree of accuracy desired, the desired permanence of installation, and the grade or fall of the ditch or stream. The degree of accuracy afforded by the various water measurement methods of course depends upon the skill of the operator as well as the proper and careful installation of the device. The generally accepted degree of accuracy using the trajectory method is ±10 percent, while orifices, flumes, and weirs can provide ± 3 percent to 5 percent, accuracy. Commercial flow meters usually fall in the range of ± 2 percent to 4 percent. The importance of proper installation and operation as well as exercising due caution when making measurements or taking readings cannot be over emphasized. Your local County Extension Director or local Soil Conservation Service District Office can provide detailed information relative to water measurement devices.

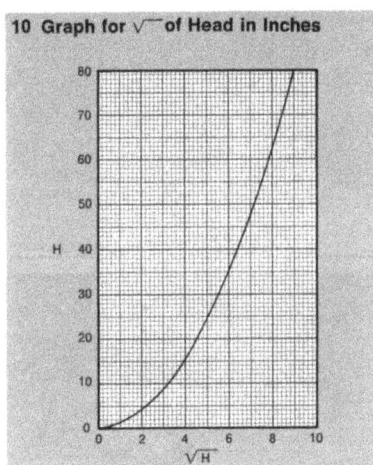

10 Graph for $\sqrt{}$ of Head in Inches

Volumetric Measurement of Irrigation Water use

The volumetric measurement of irrigation water use provides the water user with information needed to assess the performance of an irrigation system and better manage an irrigated crop. There are numerous types of volumetric measurement systems or methods that can be used to either directly measure the amount of irrigation water used or to estimate the amount of water from secondary information such as energy use, irrigation system design, or mechanical components of the irrigation system.

1) Direct Measurement Methods: Direct measurement methods usually require either the installation of a flow meter or the periodic manual measurements of flow.

Several common direct measurement systems for closed conduits (pipelines) are:

- Propeller meters.

- Orifice, venturi or differential pressure meters.

- Magnetic flux meters (both insertion and flange mount).

- Ultrasonic (travel time method).

Several common methods for direct measurement of flow in open channels are:

- Various types of weirs and flumes.

- Stage discharge rating tables.

- Area/point velocity measurements from a standardized delivery structure.

- Ultrasonic (Doppler and travel time methods).

2) Indirect Measurement Methods: Indirect measurement methods estimate the volume of water used for irrigation from the amount of energy used, irrigation equipment operating or design information, irrigation water pressure, or other information. Indirect measurements require the correlation of energy use, water pressure, system design specifications, or other parameters to the amount of water used during the irrigation or to the flow rate of the irrigation system when irrigation is occurring.

Several common indirect measurements for irrigation systems are:

- Measurement of energy used by a pump supplying water to an irrigation system.

- Measurement of end-pressure in a sprinkler irrigation system.

- Change in the elevation of water stored in an irrigation water-supply reservoir.

- Measurement of time of irrigation and size of irrigation delivery system.

Estimating irrigation water use from an indirect method can be as accurate as a direct measurement. For example, to estimate the volume of water pumped by a new electric powered irrigation pump based on kilowatt-hours of energy used during the billing period of the electric service provider, the following equation can be used:

$$\text{Acre-Feet}_{\text{Billing Period}} = \frac{\left(\left(\text{Kilowatt Hours/ Billing period}\right) \times \text{Pumping Plant Efficiency}\left(\text{percent}\right)\right)}{236.6 \times \text{Pump Pressure}\left(\text{pounds/ square inch}\left(\text{guage}\right)\right)}$$

where the Pump Pressure is the total dynamic head (feet) of the pump converted to pressure and Pumping Plant Efficiency (typically 55 to 75 percent) equals the pump efficiency (usually obtained from the pump manufacturer's pump curves, typically 60 to 80 percent) multiplied by the motor efficiency (typically 90 to 95 percent for 3-phase motors greater than 20 horsepower). The total dynamic head for a turbine pump installed in a water well includes the head required to lift the water from the well and head lost to friction.

Implementation

When implementing this best management practice it is important to be aware that the installation of a flow meter or indirect measurement varies significantly with each site, type of measurement being made, desired accuracy of the measurement, and the volume or flow rate of the water being measured. Each type of direct measurement flow meter should be installed according to the recommendations of the manufacturer of the meter. Also, some direct measurement flow methods can be unreliable or inaccurate in certain situations, particularly open channels with debris. Indirect measurement methods require the water user to determine the correlation between the indirect measurement (kilowatt hours, gallons, or cubic feet of fuel) and the volume of water used. Both direct and indirect measurement methods should be periodically evaluated for the accuracy of volume or flow rate of the water being measured.

Scope and Schedule

The methods for volumetric measurement of irrigation water and the associated scope vary from site to site, and each site and method may have unique limitations or requirements. The scope for volumetric measurement ranges from very simple (recording the amount of energy used per month from an energy bill), to complex (installation and management of a large open channel flow measurement station). Furthermore, metering requirements vary by geographic region and by political subdivision (for example, river authority, irrigation district, water improvement district, groundwater conservation district).

For direct measurement systems, the time required to install a flow meter can vary from an hour or two for a saddle mount or insertion meter to several days for the construction of a metering vault and fabrication of associated piping or the construction of a weir, flume, or open channel metering station. For indirect measurement, once the indirect measurement (such as energy usage) is correlated to the volume of water used, no additional installation or construction is required. However, the indirect measurement correlation may need to be repeated periodically to verify pumping capabilities due to normal wear on irrigation equipment.

Measuring Implementation and Determination of Water Savings

The water user should record the total quantity of water used per site, field, or system on a periodic basis as determined by the water user to be necessary for implementing other best management practices. At a minimum, recording the volume of irrigation water used should be done every year. Indirect measurements, such as energy use, are often documented by a monthly bill or statement from the supplier of the energy (that is, the electric service provider), which becomes the record of the amount of water used during such billing period.

This practice is used in coordination with other best management practices and in itself does not directly conserve any water. However, the information gained helps better inform the user of costs associated with water use and will assist the user in using water more efficiently.

Cost-effectiveness Considerations

Cost for volumetric measurement of irrigation water use varies greatly from application to application. Typical impeller meter installations for irrigation pipelines with diameters between 4 and 15 inches cost between $1,100 and $2,000 per meter. Cost for installation of a large open channel flow meter (flume, weir, or metering station) can be in the tens of thousands of dollars. Cost for indirect measurements, such as energy use, depends on the amount of time required to correlate the indirect measurement to the amount of water used and the time required to compile and record such information.

Determination of the Impact on other Resources

Because this best management practice does not directly conserve water, it does not have a direct impact on other resources. But used as a management tool that can result in water savings, energy used from pumping water is also impacted.

Water Lifting for Irrigation

Principles of Water Lifting

Energy is required, by definition, to do work; the rate at which it is used is defined as power (. A specific amount of work can be done quickly using a lot of power, or slowly using less power, but in the end the identical amount of energy is required (ignoring "side-effects" like efficiencies).

The cost of pumping or lifting water, whether in cash or kind, is closely related to the rate at which power is used (i.e. the energy requirement in a given period). Since there is often confusion on the meaning of the words "power" and "energy", it is worth also mentioning that the energy requirement consists of the product of power and time; for example, a power of say, 5kW expended over a period of say, 6h (hours), represents an energy consumption of 30kWh (kilowatt-hours). The watt (W), and kilowatt (kW) are the recommended international units of power, but units such as horsepower (hp) and footpounds per second (ft.lb/s) are still in use in some places. The joule (J) is the internationally recommended unit of energy; however it is not well known and is a very small unit, being equivalent to only 1 Ws (watt-second). For practical purposes it is common to use MJ (megajoules or millions of joules), or in the world outside scientific laboratories, kWh (kilowatt-hours). lkWh (which is one kilowatt for one hour or about the power of two horses being worked quite hard for one hour) is equal to 3.6MJ. Fuels of various kinds have their potency measured in energy terms; for example petroleum fuels such as kerosene or diesel oil have a gross energy value of about 36MJ/litre, which is almost exactly 10kWh/litre. Engines can only make effective use of a fraction of this energy, but the power of an engine will even so be related to the rate at which fuel (or energy) is consumed.

The hydraulic power required to lift or pump water is a function of both the apparent vertical height lifted and the flow rate at which water is lifted.

where $P_{hyd} = \rho g H_a$ is the hydraulic power

and ρ = density of water

g = acceleration due to gravity

H_4 = vertical height

Q = flow rate

In other words, power needs are related pro rata to the head (height water is lifted) and the flow rate. In reality, the actual pumping head imposed on a pump, or "gross working head", will be somewhat greater than the actual vertical distance, or "static head", water has to be raised. Figure indicates a typical pump installation, and it can be seen that the gross pumping head, (which determines the actual power need), consists of the sum of the friction head, the velocity head and the actual static head (or lift) on both the suction side of the pump (in the case of a pump that sucks water) and on the delivery side.

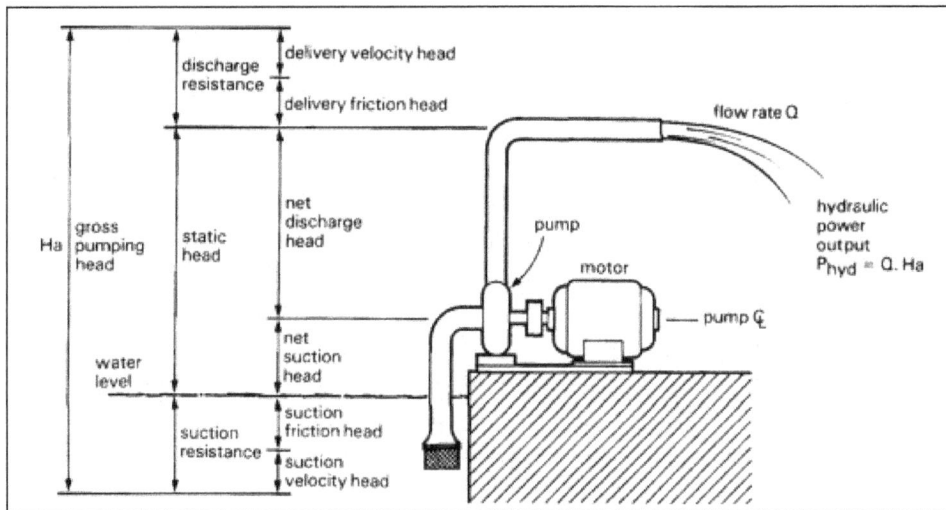

Typical pump installation.

The friction head consists of a resistance to flow caused by viscosity of the water, turbulence in the pump or pipes, etc. It can be a considerable source of inefficiency in badly implemented water distribution systems, as it is a function which is highly sensitive to flow rate, and particularly to pipe diameter, etc.

The velocity head is the apparent resistance to flow caused by accelerating the water from rest to a given velocity through the system; any object or material with mass resists any attempt to change its state of motion so that a force is needed to accelerate it from rest to its travelling velocity. This force is "felt" by the pump or lifting device as extra resistance or head. Obviously, the higher the velocity at which water is propelled through the system, the greater the acceleration required and the greater the velocity head. The velocity head is proportional to the square of the velocity of the water. Therefore, if the water is pumped out of the system as a jet, with high velocity (such as is needed for sprinkler irrigation systems), then the velocity head can represent a sizeable proportion of the power need and hence of the running costs. But in most cases where water emerges from a pipe at low velocity, the velocity head is relatively small.

Efficiency of Components: The Importance of Matching

The general principle that:

$$\text{power} \ = \ (\text{head} \times \text{flowrate})$$

and $\quad \text{energy} = (\text{head} \times \text{total weight of water lifed})$

Applies to any water lifting technique, whether it is a centrifugal pump, or a rope with a bucket on it. The actual power and energy needs are always greater then the hydraulic energy need, because losses inevitably occur when producing and transmitting power or energy due to friction. The smaller the friction losses, the higher the quality of a system. The quality of a system in terms of minimizing losses is defined as its "efficiency":

$$\text{where efficiency} \ = \frac{(\text{hydraulic energy ouput})}{(\text{actual energy input})}$$

Using energy values in the equation gives the longer-term efficiency, while power values could be used to define the instantaneous efficiency.

A truly frictionless pumping system would in theory be 100% efficient; i.e. all the energy applied to it could reappear as hydraulic energy output. However, in the real world there are always friction losses associated with every mechanical and hydraulic process. Each component of a pumping system has an efficiency (or by implication, an energy loss) associated with it; the system efficiency or total efficiency is the product of multiplying together the efficiencies of all the components. For example, a small electrically driven centrifugal pump consists of an electric motor, (efficiency typically 85%), a mechanical transmission (efficiency if direct drive of say 98%), the pump itself (optimum efficiency say 70%) and the suction and delivery pipe system (say 80% efficient). The overall system efficiency will be the product of all these component efficiencies. In other words, the hydraulic power output, measured as (static head) x (flow) (since pipe losses have been considered as a pipe system efficiency) will in this case be 47%, derived as follows:

$$0.85 \times 0.98 \times 0.7 \times 0.8 = 0.47$$

The efficiency of a component is generally not constant. There is usually an operating condition under which the efficiency is maximized or the losses are minimized as a fraction of the energy throughput; for example a centrifugal pump always has a certain speed at a given flow rate and head at which its efficiency is a maximum. Similarly, a person or draft animal also has a natural speed of operation at which the losses are minimized and pumping is easiest in relation to output.

Therefore, to obtain a pumping system which has a high overall efficiency depends very much on combining a chain of components, such as a prime-mover, transmission, pump and pipes, so that at the planned operating flow rate and static head, the components are all operating close to their optimum efficiencies - i.e. they are "well matched". A most important point to consider is that it is common for irrigation systems to perform badly even when all the components considered individually are potentially efficient, simply because one or more of them sometimes are forced to operate well away from their optimum condition for a particular application due to being wrongly matched or sized in relation to the rest of the system.

Irrigation System Losses

The complete irrigation system consists not only of a water source and water lifting mechanism and its prime-mover and energy supply, but then there must also be a water conveyance system to carry the water directly to the field or plots in a controlled manner according to the crop water requirements. There may also be a field distribution system to spread the water efficiently within each field. In some cases there could be a water storage tank to allow finite quantities of water to be supplied by gravity without running the water lifting mechanism. Figure indicates the key components of any irrigation system, and also shows some examples of common options that fulfil the requirements and which may be used in a variety of combinations.

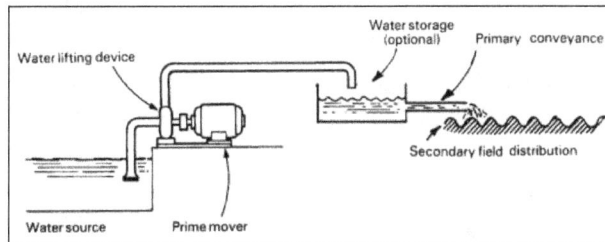

Key components of an irrigation system.

Most of the irrigation system components influence the hydraulic power requirements. For example, if pipes are used for distribution, even if they transfer water horizontally, pipe friction will create an additional resistance "felt" at the pump, which in effect will require extra power to overcome it. If open channels are used, extra power is still needed because although the water will flow freely by gravity down the channel, the input end of the channel needs to be high enough above the field to provide the necessary slope or hydraulic gradient to cause the water to flow at a sufficient rate. So the outlet from the pump to the channel needs to be slightly higher than the field level, thus requiring an increased static head and therefore an increased power demand.

For the same reason the secondary or field distribution system will also create an additional pumping head, either because of pipe friction, or if sprinklers are used then extra pressure is needed to propel the jets of water. Even open channels or furrows imply extra static head because of the need to allow for water to flow downhill.

The power needed is the product of head and flow, and any losses that cause water to fail to reach the plants also represent a reduction in effective flow from the system. Such losses therefore add to the power demand and represent a further source of inefficiency. Typical water losses are due to leakage from the conveyance system before reaching the field, evaporation and percolation into the soil away from crop roots.

Therefore, in common with the prime-mover and the water lifting device an entire irrigation system can be sub-divided into stages, each of which has a (variable) efficiency and a discrete need for power, either through adding to the actual pumping head or through decreasing the effective flow rate due to losses of water (or both).

Most components have an optimum efficiency. In the case of passive items like pipes or distribution systems this might be redefined as "cost-effectiveness" rather than mechanical efficiency. All components need to be chosen so as to be optimized close to the planned operating condition of the system if the most economical and efficient system is to be derived. The concept of "cost-effectiveness" is an important one in this connection, since most irrigation systems are a compromise or trade-off between the conflicting requirements of minimizing the capital cost of the system and minimizing the running costs. This point may be illustrated by a comparison between earth channels and aluminium irrigation pipes as a conveyance; the channels are usually cheap to build but require regular maintenance, offer more resistance to flow and, depending on the soil conditions, are prone to lose water by both percolation and evaporation. The pipe is expensive, but usually needs little or no maintenance and involves little or no loss of water.

Because purchase costs are obvious and running costs (and what causes them) are less clear, there is a tendency for small farmers to err on the side of minimizing capital costs. They also do this as they so often lack finance to invest in a better system. This frequently results in poorer irrigation system efficiencies and reduced returns then may be possible with a more capital-intensive but better optimized system.

Flow through Channels and Pipes

It is proposed here to provide an outline of the basic principles so far as they are important to the correct choice and selection of water lifting system.

Channels

When water is at rest, the water level will always be horizontal; however, if water flows down an open channel or canal, the water level will slope downwards in the direction of flow. This slope is called the "hydraulic gradient"; the greater the frictional resistance to flow the steeper it will be. Hydraulic gradient is usually measured as the ratio of the vertical drop per given length of channel; eg. lm per 100m is expressed as 1/100 or 0.01. The rate of flow (Q) that will flow down a channel depends on the cross sectional area of flow (A) and the mean velocity (v). The relationship between these factors is:

$$Q = vA$$

For example, if the cross sectional area is 0.5m2, and the mean velocity is lm/s, then the rate of flow will be:

$$1 \times 0.5 = 0.5 \, m^3 / 5$$

Table: Suggested Maximum Flow Velocities, Coefficients of Roughness and Side Slopes, for Lined and Unlined Ditches and Flumes.

Type of surface	Maximum flow velocities		Coefficients of roughness (n)	Side slopes or shape
	Metres per second	Feet per second		
UNLINED DITCHES				

Sand	0.3-0.7	1.0-2.5	0.030-0.040	3:1
Sandy loam	0.5-0.7	1.7-2.5	0.030-0.035	2:1 to 2 1/2:1
Clay loam	0.6-0.9	2.0-3.0	0.030	11/2:l to 2:1
Clays	0.9-1.5	3.0-5.0	0.025-0.030	1:1 to 11/2:1
Gravel	0.9-1.5	3.0-5.0	0.030-0.035	1:1 to 11:1
Rock	1.2-1.8	4.0-6.0	0.030-0.040	1/4:1 to 1:1
LINED DITCHES				
Concrete				
Cast-in-place	1.5-2.5	5.0-7.5	0.014	1:1 to 1 1/2 :1
Precast	1.5-2.0	5.0-7.0	0.018-0.022	11/2:1
Bricks	1.2-1.8	4.0-6.0	0.018-0.022	11/2:1
Asphalt				
Concrete	1.2-1.8	4.0-6.0	0.015	1:1 to 11/2:1
Exposed membrane	0.9-1.5	3.0-5.0	0.015	11/2:1 to 1:1
Buried membrane	0.7-1.0	2.5-3.5	0.025-0.030	2:1
Plastic				
Buried membrane	0.6-0.9	2.0-3.0	0.025-0.030	21/2:11
FLUMES				
Concrete	1.5-2.0	5.0-7.0	0.0125	
Metal				
Smooth	1.5-2.0	5.0-7.0	0.015	
Corrugated	1.2-1.8	4.0-6.0	0.021	
Wood	0.9-1.5	3.0-5.0	0.014	

The mean velocity (v) of water in a channel can be determined with reasonable accuracy for typical irrigation channels by the Chezy Formula:

$$V = C\sqrt{rs}$$

where C is the Chezy coefficient which is dependent on the roughness of the surface of the channel (n), its hydraulic radius (r), (which is the area of cross-section of submerged channel divided by its wetted submerged perimeter), and the hydraulic gradient (s) of the channel (measured in unit fall per unit length of channel).

The Chezy coefficient is found from Manning's Formula:

$$C = K\frac{r^{1/6}}{n}$$

in this formula; K = 1 if metric units are used, or K = 1.486 if feet are used; r is the previously defined hydraulic radius and n is the Manning's Coefficient of Roughness appropriate to the material used to construct the channel, Combining the above equations gives an expression for the quantity of water that will flow down a channel under gravity as follows:

$$Q = AK \frac{r^{2/3} s^{1/2}}{n}$$

where Q will be in m³/s, if A is in m², r is in metres, and K is 1.

To obtain a greater flow rate, either the channel needs to be large in cross section (and hence expensive in terms of materials, construction costs and land utilization) or it needs to have a greater slope. Therefore irrigation channel design always introduces the classic problem of determining the best trade-off between capital cost or first cost (i.e. construction cost) and running cost in terms of the extra energy requirement if flow is obtained by increasing the hydraulic gradient rather than the cross sectional area. The nature of the terrain also comes into consideration, as channels normally need to follow the natural slope of the ground if extensive regrading or supporting structures are to be avoided.

Obviously in reality, the design of a system is complicated by bends, junctions, changes in section, slope or surface, etc.

A further point to be considered with channels is the likely loss of water between the point of entry to the channel and the point of discharge caused by seepage through the channel walls and also by evaporation from the open surface. Any such losses need to be made up by extra inputs of water, which in turn require extra pumping power (and energy) in proportion. Seepage losses are of course most significant where the channel is unlined or has fissures which can lose water, while evaporation only becomes a problem for small and medium scale irrigation schemes with channels having a large surface area to depth ratio and low flow rates, particularly under hot and dry conditions; the greatest losses of this kind occur generally within the field distribution system rather than in conveying water to the field. The main factors effecting the seepage rate from a channel or canal are:

- Soil characteristics.

- Depth of water in the channel in relation to the wetted area and the depth of the ground-water.

- sediment in water in relation to flow velocity and length of time channel has been in use.

This latter point is important , as any channel will leak much more when it has been allowed to dry out and then refill. Seepage decreases steadily through the season due to sediment filling the pores and cracks in the soil. Therefore, it is desirable to avoid letting channels dry out completely to reduce water losses when irrigating on a cyclic basis.

Typical conveyance efficiencies for channels range at best from about 90% (or more) with a heavy clay surface or a lined channel in continuous use on small to medium land holdings down to 60-80% in the same situation, but with intermittent use of the channel.

In less favourable conditions, such as on a sandy or loamy soil, also with intermittent use, the conveyance efficiency may typically be 50-60% or less; (i.e. almost half the water entering the channel failing to arrive at the other end).

Methods for calculating conveyance losses have been derived and are discussed in detail in specialist references (such as). For example, an approach used by the Irrigation Department in Egypt uses an empirical formula attributed to Molesworth and Yennidumia:

- $S = c\,L\,P\sqrt{R}$

where, S will be the conveyance loss in m3/s per length L

- if
    c = coefficient depending on nature of soil
    $(eg.\, c = 0.0015\text{ for clay and}.003\text{ for sand})$
    L = length in Km
    P = wetted perimrter of cross section
    R = hydraulic mean depth $(ie.\text{ flow cross-sectional area divided by width of surface})$

Pipes

A pipe can operate like a channel with a roof on it; i.e. it can be unpressurized, often with water not filling it. The advantage of a pipe, however, is that it need not follow the hydraulic gradient like a channel, since water cannot overflow from it if it dips below the natural level. In other words, although pipes are more expensive than channels in relation to their carrying capacity, they generally do not require accurate levelling and grading and are therefore more cheaply and simply installed. They are of course essential to convey water to a higher level or across uneven terrain. As with a channel, a pipe also is subject to a hydraulic gradient which also necessarily becomes steeper if the flow is increased; in other words a higher head or higher pressure is needed to overcome the increased resistance to a higher flow. This can be clarified by imagining a pipeline with vertical tappings of it. When no flow takes place due to the outlet valve being closed, the water pressure along the pipe will be uniform and the levels in the vertical tappings will correspond to the head of the supply reservoir. If the valve is opened so that water starts to flow, then a hydraulic gradient will be introduced as indicated in the second diagram and the levels in the vertical tappings will relate to the hydraulic gradient, in becoming progressively lower further along the pipe. The same applies if a pump is used to push water along a pipe as in the lowest diagram in the figure. Here the pump needs to overcome a resistance equal to the static head of the reservoir indicated in the two upper diagrams, which is the pipe friction head. In low lift applications, as indicated, the pipe friction head can in some cases be as large or larger then the static head (which in the example is all suction head since the pump is mounted at the same level as the discharge). The power demand, and hence the energy costs will generally be directly related to total head for a given flow rate, so that in the example, friction losses in the pipe could be responsible for about half the energy costs.

An approximate value of the head loss through a pipe can be gained using the empirical equation:

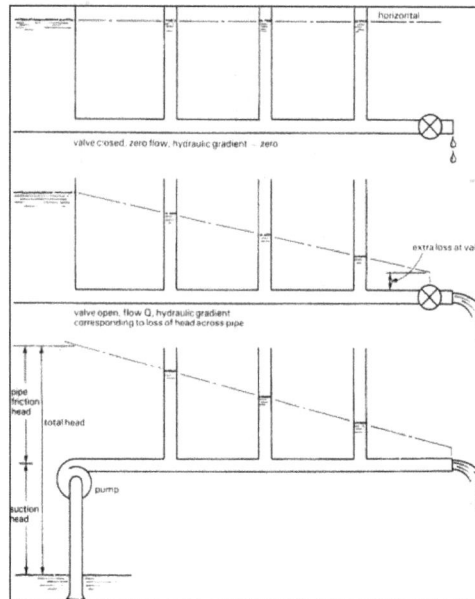

The concept of an 'hydraulic gradient'.

$$H_f = K \frac{LQ^2}{C^2 D^5}$$

L = length of pipe

Q = flow rate

C = coefficient of friction for pipe

D = internal diameter of pipe

The head loss due to friction is expressed as an "hydraulic gradient", i.e. head per length of pipe (m per m or ft per ft).

Note: use K = 10 with metric units, (L and D in metres and Q in cubic metres per second), and K = 4.3 with L and D in feet and Q in cubic feet per second. Values of C are typically 1.0 for steel, 1.5 for concrete, 0.8 for plastics.

An easy way to estimate pipe friction is to use charts, such as figure. Reference to this figure indicates that a flow, for example, of 6 litres/second (95 US gall/minute) through a pipe of 80mm (3" nominal bore) diameter results in a loss of head per 100m of pipe of just over 2m., As an alternative method, figure gives a nomogram (from reference) for obtaining the head loss, given in this case as m/km, for rigid PVC pipe. These results must be modified, depending on the type of pipe, by multiplying the result obtained from the chart by the roughness coefficient of the pipe relative to the material for which the chart of nomogram was derived; for example, if figure is to be used for PVC pipe, the result must be multiplied by the factor 0.8 (as indicated at the foot of the figure) because PVC is smoother than iron and typically therefore imposes only 80% as much friction head.

Account must also be taken of the effects of changes of cross section, bends, valves or junctions, which all tend to create turbulence which in effect raises the effective friction head. Ageing of pipes due to growth of either organic matter or corrosion, or both, also increases the friction head per unit length because it increases the frictional resistance and it also decreases the available cross section of flow. This is a complex subject and various formulae are given in text books to allow this effect to be estimated when calculating head losses in pipes.

The head loss due to friction in a pipeline is approximately related to the mean velocity and hence the flow rate squared; i.e.:

$$\text{head loss } h_f = KQ_2$$

therefore, the total head felt by a pump will be approximately the sum of the static head, the friction head and (if the water emerges from the outlet with significant velocity) the velocity head:

$$\text{total head } H_t = h_s + KQ_2 + \frac{V^2}{2g}$$

i.e. (total head) = (static head) + (friction head) + (velocity head)

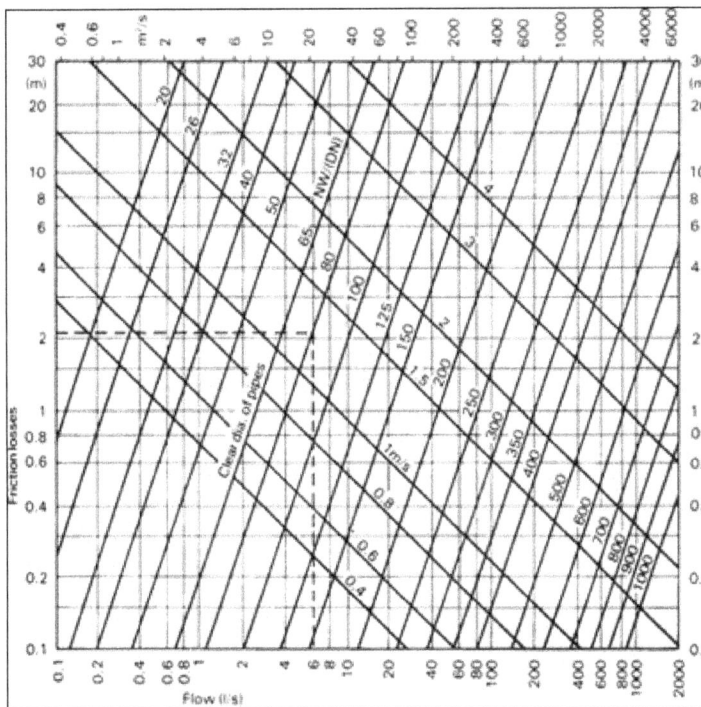

Friction losses in meters per 100m for a new pipeline of cast iron.

For other types of pipe multiply the friction loss as indicated by the table by the factors given below:

New Rolled steel	-	0.8
New Plastic	-	0.8
Old rusty cast iron	-	1.25
Pipes with encrustations	-	1.7

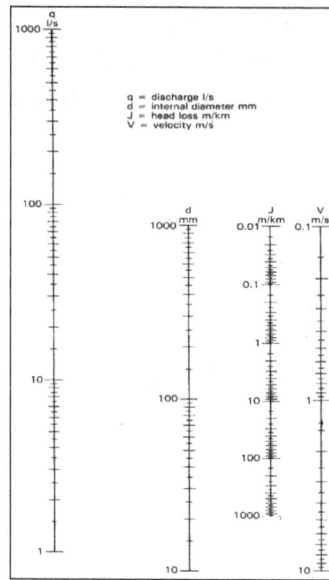

Determination of head friction losses in straight pipes.

$$\text{total head } H_t = h_s + K'Q_2$$

Head loss nomogram calculated for rigid PVC pipes using Blasius formula.

Since the velocity of flow is proportional to the flow rate (Q), the above equation can be re-written:

where,

$$K' = K + \frac{1}{2gA^2}$$

Figure illustrates the relationship between the total head and the flow rate for a pumped pipeline, and the pipeline efficiency which can be expressed in energy terms as:

$$\text{pipeline efficiency } \eta_{\text{pipe}} = \frac{h_s - K'Q^2}{h_s}$$

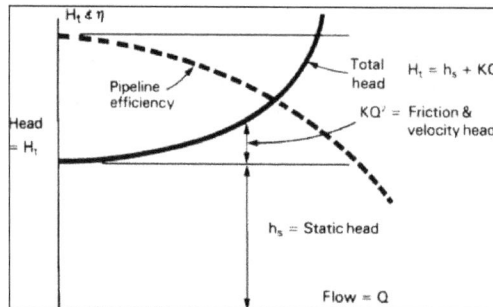

Pipeline and efficiency vary with flow.

Suction Lift: the Atmospheric Limit

Certain types of pump are capable of sucking water from a source; i.e. the pump can be located above the water level and will literally pull water up by creating a vacuum in the suction pipe.

Drawing water by suction depends on the difference between the atmospheric pressure on the free surface of the water and the reduced pressure in the suction pipe developed by the pump. The greater the difference in pressure, the higher the water will rise in the pipe. However, the maximum pressure difference that can be created is between sea level atmospheric pressure on the free surface and a pure vacuum, which theoretically will cause a difference of level of water of 10.4m (or 34ft). However, before a drop in pressure even approaching a pure vacuum can be produced, the water will start gassing due to release of air held in solution (just like soda water gasses when released from a pressurized container); if the pressure is reduced further, the water can boil at ambient temperature. As soon as this happens, the pump loses its prime and the discharge will cease (due to loss of prime) or at least be severely reduced. In addition, boiling and gassing within the pump (known as cavitation) can cause damage if allowed to continue for any length of time.

The suction lifts that can be achieved in practice are therefore much less than 10.4m. For example, centrifugal pumps, which are prone to cavitation due to the high speed of the water through the impeller, are generally limited to a suction lift of around 4.5m (15ft) even at sea level with a short suction pipe. Reciprocating pumps generally impose lower velocities on the water and can therefore pull a higher suction lift, but again, for practical applications, this should never normally exceed about 6.5m (21ft) even under cool sea level conditions with a short suction pipe.

At higher altitudes, or if the water is warmer than normal, the suction lift will be reduced further. For example, at an altitude of 3 000m (10 000ft) above sea level, due to reduced atmospheric pressure, the practical suction lift will be reduced by about 3m compared with sea level, (and proportionately for intermediate altitudes, so that 1 500m above sea level will reduce suction lift by about 1.5m). Higher water temperatures also cause a reduction in practical suction head; for example, if the water is at say 30°C, (or 86°F) the reduction in suction head compared with water at a more normal 20°C will be about 7%.

Extending the length of the suction pipe also reduces the suction head that is permissible, because pipe friction adds to the suction required; this effect depends on the pipe diameter, but typically a suction pipe of say 80m length will only function satisfactorily on half the above suction head.

Draw-down and Seasonal Variations of Water Level

Groundwater and river water levels vary, both seasonally and in some cases due to the rate of pumping. Such changes in head can significantly influence the power requirements, and hence the running costs. However, changes in head can also influence the efficiency with which the system works, and thereby can compound any extra running costs caused by a head increase. More serious problems can arise, resulting in total system failure, if for example a surface mounted suction pump is in use, and the supply water level falls sufficiently to make the suction lift exceed the practical suction lift limits.

Figure illustrates various effects on the water level of a well in a confined aquifer. The figure shows that there is a natural ground water level (the water table), which often rises either side of a river or pond since ground water must flow slightly downhill into the open water area. The water table tends to develop a greater slope in impermeable soils (due to higher resistance to flow and greater capillary effects), and is fairly level in porous soil or sand.

If a well is bored to below the water table and water is extracted, the level in the well tends to drop until the inflow of water flowing "downhill" from the surrounding water table balances the rate at which water is being extracted. This forms a "cone of depression" of the water table surrounding the well. The greater the rate of extraction, the greater the drop in level. The actual drop in level in a given well depends on a number of factors, including soil permeability and type, and the wetted surface area of well below the water table (the greater the internal surface of the well the greater the inflow rate that is possible). Extra inflow can be gained either by increasing the well diameter (in the case of a hand-dug well) or by deepening it (the best possibility being with a bore-hole).

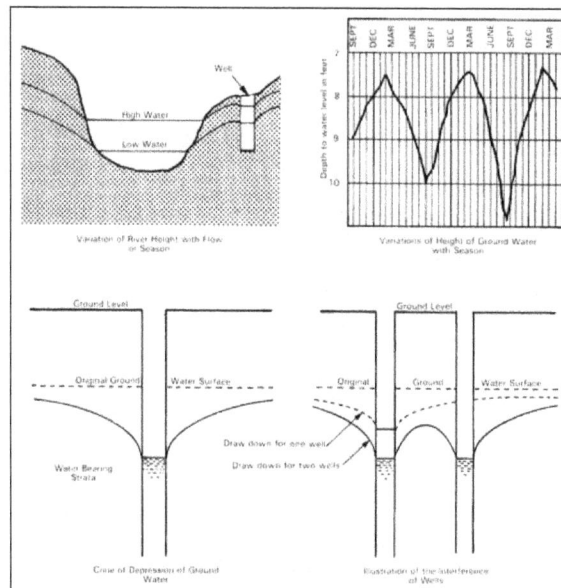

Effects of various physical conditions on the elevation of water surfaces in wells.

Draw-down usually will increase in proportion to extraction rate. A danger therefore if large and powerful pumps are used on small wells or boreholes is to draw the water down to the pump intake level, at which stage the pump goes on "snore" (to use a commonly used descriptive term). In other words, it draws a mixture of air and water which in many cases causes it to lose its prime and cease to deliver. As with cavitation, a "snoring" pump can soon be damaged. But not only the pump is at risk; excessive extraction rates on boreholes can damage the internal surface below the water table and cause voids to be formed which then leads to eventual collapse of the bore. Even when a fully lined and screened borehole is used, excessive extraction rates can pull a lot of silt and other fine material out with the water and block the screen and the natural voids in the surrounding sub-soil, thereby increasing the draw-down further and putting an increasing strain on the lowermost part of the bore. Alternatively, with certain sub-soils, the screen slots can be eroded by particles suspended in the water, when the extraction rate is too high, allowing larger particles to enter the bore and eventually the possible collapse of the screen.

Neighbouring wells or boreholes can influence each other if they are close enough for their respective cones of depression to overlap, as indicated in figure. Similarly, the level of rivers and lakes will often vary seasonally, particularly in most tropical countries having distinct monsoon type seasons with most rain in just a few months of the year. The water table level will also be influenced by seasonal rainfall, particularly in proximity to rivers or lakes with varying levels.

Therefore, when using boreholes, the pump intake is best located safely below the lowest likely water level, allowing for seasonal changes and draw-down, but above the screen in order to avoid producing high water velocities at the screen.

When specifying a mechanized pumping system, it is therefore most important to be certain of the minimum and maximum levels if a surface water source is to be used, or when using a well or borehole, the draw-down to be expected at the proposed extraction rate. A pumping test is necessary to determine the draw-down in wells and boreholes; this is normally done by extracting water with a portable engine-pump, and measuring the drop in level at various pumping rates after the level has stablized. In many countries, boreholes are normally pumped as a matter of routine to test their draw-down and the information from the pumping test is commonly logged and filed in the official records and can be referred to later by potential users.

Review of a Complete Lift Irrigation System

The factors that impose a power load on a pump or water lifting device are clearly more complicated than simply multiplying the static head between the water source and the field by the flow rate. The load consists mainly of various resistances to flow which when added together comprise the gross pumping head, but it also is increased by the need to pump extra water to make up for losses between the water source and the crop.

The system hydraulic efficiency can be defined as the ratio of hydraulic energy to raise the water delivered to the field through the static head, to the hydraulic energy actually needed for the amount of water drawn by the pump:

$$\text{system hydraulic efficiency } \eta_{sys} = \frac{E_{stat}}{E_{gross}}$$

Where E_{stat} is the hydraulic energy output, and E_{gross} is hydraulic energy actually applied. Finally, figure. indicates the energy flow through typical complete irrigation water lifting and distribution systems and shows the various losses.

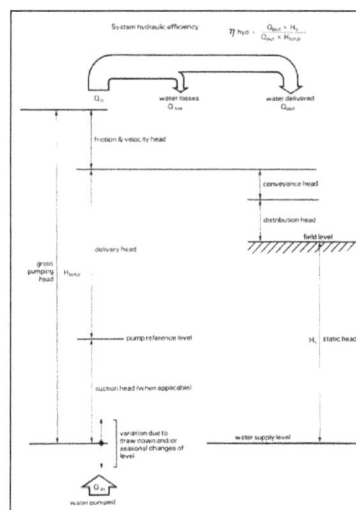

Factors affecting system hydraulic efficiency.

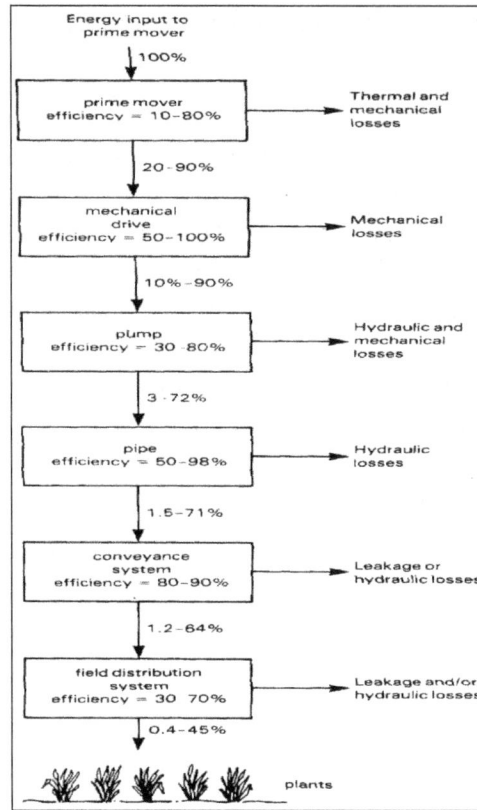

Energy flow through typical irrigation system (showing percentage of original energy flow that is transmitted from each component to the next).

Practical Power Requirements

Calculating the power requirement for water lifting is fundamental to determining the type and size of equipment that should be used, so it is worth detailing the principles for calculating it. In general the maximum power required will simply be:

$$\frac{(\text{Maximum mass flow delivered}) \times (\text{satic head}) \times g}{(\text{total system efficiency at max. flow})}$$

where the mass flow is measured in kg/s of water. 1kg of water is equal to 1 litre in volume, so it is numerically equal to the flow in litres per second; g is the acceleration due to gravity of 9.81m/s2 (or 32.2 ft/s2). Therefore, for example, 5 litre/sec through 10m with a system having an overall efficiency of 10% requires:

$$\frac{5(\text{kg/s}) \times 10(\text{m}) \times 9.81(\text{m/s}^2)}{0.10(\text{efficiency})} = 4905\,\text{w}$$

The daily energy requirement will similarly be:

$$\frac{(\text{mass of water delivered per day}) \times (\text{head}) \times g}{(\text{average system efficiency})}$$

eg. for 60m3/day lifted through 6m with an average efficiency of 5%

$$\frac{60\,000 \times 6 \times 9.81}{0.05} = 70.6\,\text{MJ/day (million joules/ day)}$$

Note: 60m3 = 60 000 litres which in turn has a mass of 60 000kg (= 60 tonne). Also, since 1kWh = 3.6MJ, we can express the above result in kWh simply by dividing by 3.6:

so,

$$70.6\,\text{MJ/day} = 19.6\,\text{KWh/day}$$

It follows from these relationships that a simple formula can be derived for converting an hydraulic energy requirement into kWh, as follows:

$$E_{hyd} = \frac{QH}{367}$$

If the above calculation relates to a gasoline engine pump irrigation system, as it might with the figures chosen for the example, then we know that as the energy input is 19.6kWh/day and as gasoline typically has an energy content of 32MJ/litre or 8.9kWh/litre, this system will typically require an input of 2.2 litre of gasoline per day.

Figure illustrates the hydraulic power requirement to lift water at a range of pumping rates appropriate to the small to medium sized land-holdings this publication relates to. These figures are the hydraulic output power and need to be divided by the pumping system efficiency to arrive at the input power requirement. For example, if a pump of 50% efficiency is used, then a shaft power of twice the hydraulic power requirement is needed. The small table on figure indicates the typical hydraulic power output of various prime movers when working with a 50% efficient water lifting device; i.e. it shows about half the "shaft power" capability. The ranges as indicated are meant to show "typical" applications; obviously there are exceptions.

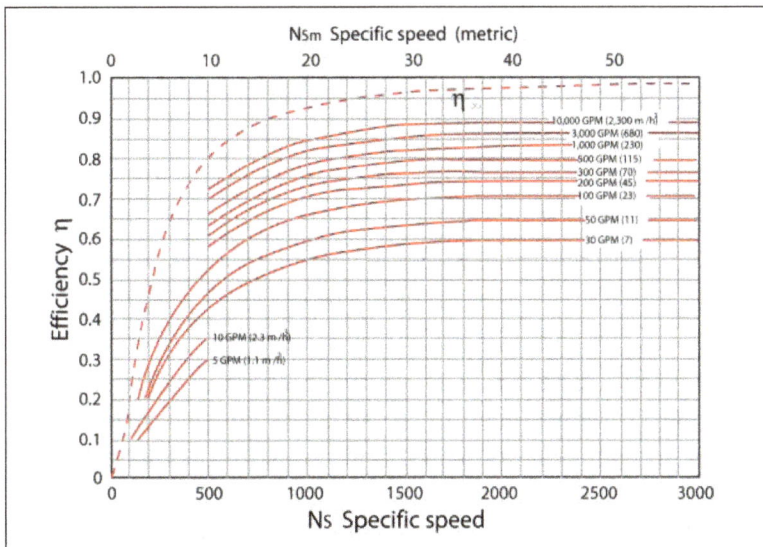

Hydraulic power requirements to lift water.

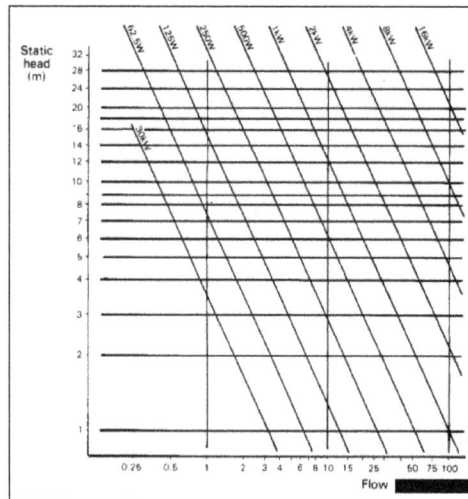

Relationship between power, head and flow.

These power curves, which are hyperbolas, make it difficult to show the entire power range of possible interest in connection with land-holdings from less than 1ha to 25ha, even though they cover the flow, head and power range of most general interest. Figure is a log-log graph of head versus flow, which straightens out the power curves and allows easier estimation of the hydraulic power requirement for flows up to 100 l/s and hydraulic powers of up to 16kW.

Figure is perhaps more generally useful, being a similar log-log graph, but of daily hydraulic energy requirement to deliver different volumes of water through a range of heads of up to 32m. The area of land that can be covered, as an example to 8mm depth, using a given hydraulic energy output over the range of heads is also given.

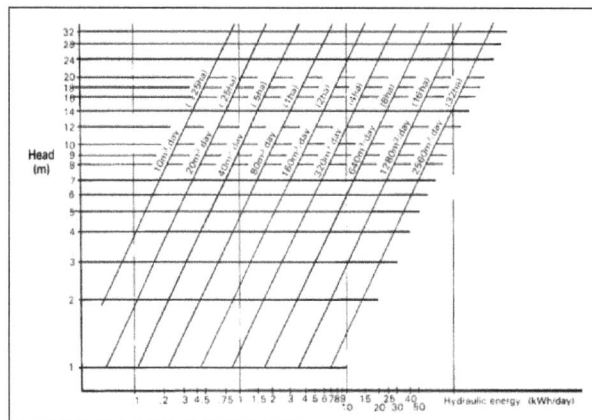

Relationship between energy, head and daily output
(areas that can be irrigated to a depth of 8mm are shown in parentheses).

Finally, figure is a nomograph which allows the entire procedure of calculating power needs for a given irrigation requirement to be reduced to ruling a few lines so as to arrive at an answer. The following example of the procedure is indicated and helps to illustrate the process; starting with the area to be irrigated (in the example 3ha is used), rule a line vertically upwards until it intersects the diagonal. This point of intersection gives the required depth of irrigation; 8mm is used in the example but field and distribution losses are not accounted for in this nomograph, so the irrigation demand used must be the gross and not the nett requirement. Rule horizontally from the point of intersection, across the vertical axis (which indicates the daily water requirement in cubic metres per day - 240 in the example) until the line intersects the diagonal relating to the pumping head; 10 m head is used in the example. Dropping a vertical line from the point of intersection gives the hydraulic energy requirement (6.5kWh (hyd)/day). This is converted to a shaft energy requirement by continuing the line downwards to the diagonal which corresponds with the expected pumping efficiency; 50% efficiency is assumed for the example (the actual figure depends on the type of pumping system) and this gives a shaft power requirement of 13kWh/day when a line is ruled horizontally through the shaft power axis. The final decision is the time per day which is to be spent pumping the required quantity of water; 5h is used as the example. Hence, ruling a line vertically from the point of intersection to the average power axis (which coincides with the starting axis), shows that a mean power requirement (shaft power) of about 2.6kW is necessary for the duty chosen in the example. It should be noted that this is mean shaft power; a significantly higher peak power or rated power may be necessary to achieve this mean power for the number of hours necessary.

This nomograph readily allows the reader to explore the implications of varying these parameters; in the example it is perhaps interesting to explore the implications of completing the pumping in say 3h rather than 5h and it is clear that the mean power requirement then goes up to about 4.25kW.

In some cases it may be useful to work backwards around the nomograph to see what a power unit of a certain size is capable of doing in terms of areas and depths of irrigation.

The nomograph has been drawn to cover the range from 0 - 10ha, which makes it difficult to see clearly what the answers are for very small land-holdings of under 1ha. However, the nomograph also works if you divide the area scale by 10, in which case it is also necessary to divide the answer in terms of power needed by 10. In the example, if we were interested in 0.3ha instead of 3ha, and if the same assumptions are used on depth of irrigation, pumping head, pump efficiency and hours per day for pumping, the result will be 0.26kW (or 260W) instead of 2.6kW as indicated. Obviously the daily water requirement from the top axis will also need to be divided by ten, and in the example will be 24m3/day. Similarly, it is possible to scale the nomograph up by a factor of ten to look at the requirements for 10 to 100 ha in exactly the same way. Note that in most real cases, if the scale is changed, factors like the pump efficiency ought to be changed too. An efficiency of 50% used in the example is a poorish efficiency for a pump large enough to deliver 240m3/day, but it is rather a high efficiency for a pump capable of only one tenth of this daily discharge.

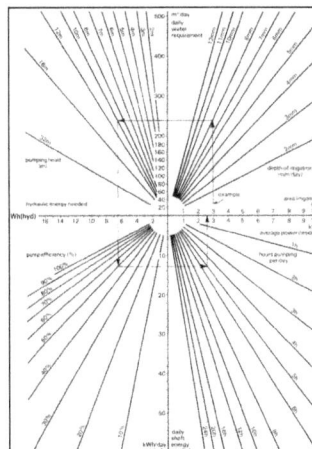

Nomogram for calculating power needs for a given area, depth of irrigation and head.

Outline of Principles of Small Scale Irrigation

Irrigation Water Requirements

The quantity of water needed to irrigate a given land area depends on numerous factors, the most important being:

- Nature of crop

- Crop growth cycle

- Climatic conditions

- Type and condition of soil

- Topography

- Conveyance efficiency

- Field application efficiency

- Water quality

- Effectiveness of water management.

Few of these factors remain constant, so that the quantity of water required will vary from day to day, and particularly from one season to the next. The selection of a small scale irrigation system needs to take all of the above factors into account.

The crop takes its water from moisture held in the soil in the root zone. The soil therefore effectively acts as a water storage for the plants, and the soil moisture needs replenishing before the moisture level falls to what is known as the "Permanent Wilting Point" where irreversible damage to the crop can occur. The maximum capacity of the soil for water is when the soil is "saturated", although certain crops do not tolerate water- logged soil and in any case this can be a wasteful use of water. In all cases there is an optimum soil moisture level at which plant growth is maximized. The art of efficient irrigation is to try to keep the moisture level in the soil as close to the optimum as possible.

Nett Irrigation Requirement

The estimation of irrigation water requirements starts with the water needs of the crop. First the "Reference Crop Evapotranspiration" ET is determined; this is a standardized rate of evapotranspiration (related to a reference crop of tall green grass completely shading the ground and not short of water) which provides a base-line and which depends on climatic factors including pan evaporation data and windspeed. A full description on the determination of ET_o is presented in reference. Because ET_o depends on climatic factors, it varies from month to month, often by a factor of 2 or more. The evapotranspiration of a particular crop (ET_{crop}) will of course be different from that of the reference crop, and this is determined from the relationship:

$$ET_{crop} = ET_o \times K_c$$

K_c is a "crop coefficient" which depends on the type of crop, its stage of growth, the growing season and the prevailing climatic conditions. It can vary typically from around 0.3 during initial growth to around 1.0 (or a bit over 1.0) during the mid-season maximum rate of growth period; shows an example. Therefore the actual value of the crop water requirement, ETcrop usually varies considerably through the growing season.

The actual nett irrigation requirement at any time is the crop evapotranspiration demand, minus any contributions from rainfall, groundwater or stored moisture in the soil. Since not all rainfall will reach the plant roots, because a proportion will be lost through run-off, deep percolation and evaporation, the rainfall is factored to arrive at a figure for "effective rainfall". Also, some crops require water for soil preparation, particularly for example, rice, and this need has to be allowed for in addition to the nett irrigation requirement.

To give an idea of what these translate into in terms of actual water requirements an approximate "typical" nett irrigation requirement under tropical conditions with a reasonably efficient irrigation system and good water management is 4 000m³/ha per crop, but under less favourable conditions as much as 13 000m³/ha per crop can be needed. This is equal to 400-1300mm of water per crop respectively. Since typical growing cycles are in the range of 100-150 days in the tropics, the average daily requirement will therefore be in the 30-130m³/ha range (3-13mm/day). Because the water demand varies through the growing season, the peak requirement can be more than double the average, implying that a nett peak output of 50-200m³/ha will generally be required (which gives an indication of the capacity of pumping system needed for a given area of field).

Inadequate applications of irrigation water will not generally kill a crop, but are more likely to result in reduced yield. Conversely, excessive applications of water can also be counterproductive apart from being a waste of water and pumping energy. Accurate application is therefore of importance mainly to maximise crop yields and to get the best efficiency from an irrigation system.

Gross Irrigation Requirement

The output from the water lifting device has to be increased to allow for conveyance and field losses; this amount is the gross irrigation requirement. Typical conveyance and field distribution system efficiencies are given in table from which it can be seen that conveyance efficiencies fall into the range 65-90% (depending on the type of system), while "farm ditch efficiency" or field application efficiency will typically be 55- 90%. Therefore, the overall irrigation system efficiency, after the discharge from the water lifting device, will be the product of these two; typically 30-80%. This implies a gross irrigation water requirement at best about 25% greater than the nett requirement for the crop, and at worst 300% or more.

The previous "typical peak nett irrigation" figures of 50-200m³/day per hectare imply "peak gross irrigation" requirements of 60-600m3/day; a wide variation due to compounding so many variable parameters. Clearly there is often much scope for conservation of pumping energy by improving the water distribution efficiency; investment in a better conveyance and field distribution system will frequently pay back faster than investment in improved pumping capacity and will achieve the same result.

Certainly costly pumping systems should generally only be considered in conjunction with efficient conveyance and field distribution techniques. The only real justification for extravagant water losses is where pumping costs are low and water distribution equipment is expensive.

Table: Average Conveyance Efficiency

Irrigation Method	Method of Water Delivery	Irrigated Area (ha)	Efficiency (%)
Basin for rice cultivation	Continuous supply with no substantial change in flow	—	90
Surface irrigation (Basin, Border and Furrow	Rotational supply based on predetermined schedule with effective management	3,000-5,000	88
	Rotational supply based on predetermined schedule with less effective management	< 1,000	70
		> 10,000	
	Rotational supply based on advance request	< 1,000	65
		> 10,000	

Table: Average Farm Ditch Efficiency

Irrigation Method	Method of Delivery	Soil Type and Ditch Condition	Block Size (ha)	Efficiency (%)
Basin for rice	Continuous	Unlined: Clay to heavy clay Lined or piped	up to 3	90
Surface	Rotation or	Unlined: Clay to heavy clay	<20	80
Irrigation	Intermittent	Lined or piped	>20	90
	Rotation or	Unlined: Silt clay	<20	60-70
	Intermittent	Lined or piped	>20	80
	Rotation or	Unlined: Sand, loam	<20	55
	Intermittent	Lined or piped	>20	65

Table: Average Application Efficiency

Irrigation Method	Method of Delivery	Soil type	Depth of Application (mm)	Efficiency (%)
Basin	Continuous	Clay	>60	40-50
		Heavy clay		
Furrow	Intermittent	Light soil	>60	60
Border	Intermittent	Light soil	>60	60
Basin	Intermittent	All soil	>60	60

A summary of the procedure so far outlined to arrive at the gross crop irrigation water requirement is given in figure.

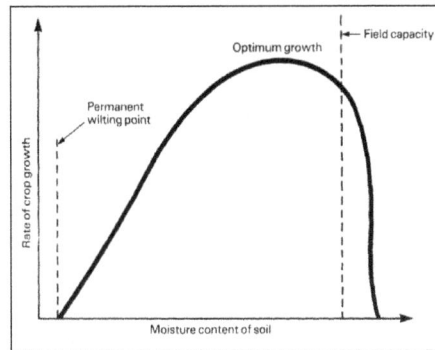

Rate of crop growth as a function of soil moisture content.

Pumping Requirement

In order to specify a water lifting system the following basic information is needed:

- The average water demand through the growing season.

- The peak daily water demand (which generally will occur when the crop coefficient and rate of plant growth are at their peak).

Having determined the daily application required by the plants, a further consideration is the "intake rate" as different soil types absorb water at different rates. Too rapid a rate of application on some soils can cause flooding and possible loss of water through run-off. This constraint determines the maximum flow rate that can usefully be absorbed by the field distribution system. For example, some silty clay soils can only take about 7 l/sec per hectare, but in contrast sandy soils do not impose a serious constraint as they can often usefully absorb over 100 l/s per hectare. Obviously lower rates than the maximum are acceptable, although the application efficiency is likely to be best at a reasonably high rate in most cases, and farmers obviously will prefer not to take longer than necessary to complete the job.

Taking account of the above constraint on flow rate, it is then possible to calculate how many hours per day the field will require irrigating.

Table: Average Intake Rates of Water in mm/hr for Different Soils and Corresponding Stream Size 1/sec/ha.

Soil Texture	Intake Rate mm/hr	Stream size q 1/sec/ha
Sand	50 (25 to 250)	140
Sandy loam	25 (15 to 75)	70
Loam	12.5 (8 to 20)	35
Clay loam	8 (2.5 to 5)	7
Silty clay	2.5 (0.03 to 5)	7
Clay	5 (1 to 15)	14

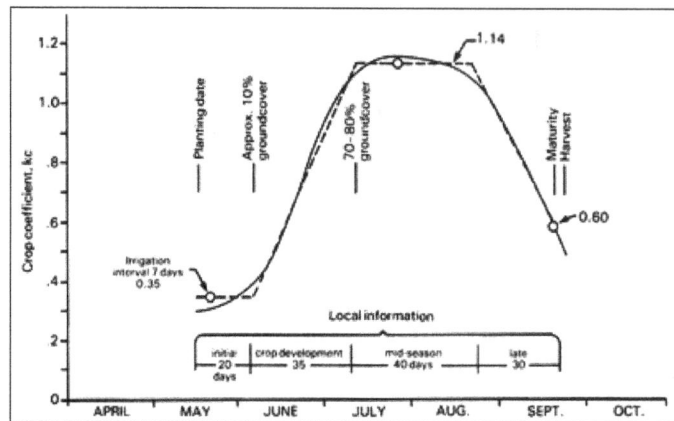

Example of a crop coefficient curve for corn planted in mid-May at Cairo,
Egypt; e.g. initial stage is 8.4mm/day with irrigation.

Hydraulics of Drip Irrigation System

The pipes of drip irrigation system are made of plastics and comprise of the main line, sub-mains and laterals. Drip irrigation system design must ensure nearly uniform discharge of the drippers in each section that is controlled by a valve and irrigated as unit of the system. The maximum pressure difference allowable in a system is 20% and the maximum difference in pressure between the head end and the tail end of a lateral should not exceed 10%. The relationship between pressure and discharge for different types of emission devices can be obtained from the manufacturers catalogues. The pressure loss can be estimated from monographs tables or using the relationships expressed in the form of equations. Head loss occurs due to friction between the pipe walls and water as it flows through the system. Obstacles- turns, bends, expansions, contractions of pipes, etc., along the way to flow increase head losses.

The head loss due to friction is a function of the following variables:

1. Pipe length.

2. Pipe diameter.

3. Pipe wall smoothness.

4. Water flow rate.

5. Liquid viscosity.

Pressure Variation in Irrigation Pipe Line

The major requirement in most situations is that the irrigation system must apply water uniformly over the entire field. The performance of the drip system is related to operating pressure. The uniform water application from a drip emitter requires required desired optimum pressure. Friction loss in pipes and fittings, and differences in elevation cause pressure to vary in a field. Friction loss causes the pressure to decrease in the downstream direction, while changes in elevation can cause either an increase or decrease in pressure due to pipe running on uphill or downhill. The difference in pressure between locations along pipe line can be estimated as:

$$P_d = P_u - 9.81\left(h_l \pm_\Delta Z\right)$$

Where,

P_d & P_u = pressure at down and upstream positions, respectively, kPa

h_l = energy loss in pipe between the up- and downstream positions, m

ΔZ = elevation difference, m (+ve for uphill & -ve for downhill

The energy loss (h_l) includes head loss due to friction and minor loss, which can be estimated as:

$$h_l = FH_f + M_l$$

where,

F = constant; f (number of outlets and method used to estimate, H_f)

H_f = friction loss in pipe between up and downstream locations, m

M_l = minor losses through fittings, m

Major and minor losses are two types of losses that occur in pipe flow.

Major Losses

Major losses occur while water flow along straight pipes. The universal equation used to calculate friction losses of water flow along a pipe is known as the Hazen-Williams equation, given by,

$$H_f\left(100\right) = K\left(\frac{Q}{C}\right)^{1.852} \times D^{-4.871} \times F$$

As the length of the pipe increases, the discharge in the pipe decreases due to emission outlets and hence the total energy drop is less than as estimated by the above equation. For this reason, a reduction factor F is introduced.

where,

$H_f(100)$ = head loss due to friction per 100 meter of pipe length, m/100m

K = a constant which is 1.22×10^{12} in metric units

D = inner pipe diameter, mm

C = friction coefficient (indicates inner pipe wall smoothness, the higher the C coefficient, the lower the head loss)

Q = flow rate, L s^{-1}

The Hazen-Williams equation is valid in a limited range of temperature and flow pattern. In small diameter laterals, the Darcy-Weisbach equation gives better results in calculating head loss due to friction in small diameter lateral pipes. It is given by,

$$H_f = f\left(\frac{LV^2}{2Dg}\right)$$

where,

H_f = head loss, m L = pipe length, m

f = Darcy-Weisbach friction factor V = Flow velocity, m s^{-1}

g = gravitation acceleration (9.81, m s^{-2}) D = inner pipe diameter, m

Both the Hazen-Williams and Darcy-Weisbach equations include a parameter for the smoothness of the internal surface of the pipe wall. In Hazen-Williams, it is the dimensionless C coefficient and with Darcy-Weisbach the roughness factor f, as the C coefficient is higher, head loss will be lower. On the opposite, in the Darcy-Weisbach equation, higher values of indicate higher head losses.

Minor Losses

The minor losses through fittings can be estimated or obtained from standard tables available in text books and hydraulics manuals/ hand books. Minor losses are created by the flow at bends and transitions. If the flow velocities are high through many bends and transitions in the system, minor losses can build up and become substantial losses. Minor head losses are expressed as an equivalent length factor that adds a virtual length of straight pipe of the accessory diameter to the length of the pipe under calculation.

The Darcy-Weisbach, Hazen Williams or Scobey equation can be used to compute head loss due to friction, H_f. The general form of these equations can be written as:

$$H_f = \frac{(K)(c)(L)(Q^m)}{D^{2m+n}}$$

where,

K = friction factor that depends on pipe material

L = length of pipe, m

Q = flow rate, L min^{-1}

D = diameter of pipe, mm

c, m, n = constants can be obtained from table.

Table: Constants of friction loss equations

Equations for computing H_f	c	m	n
Darcy-Weisbach	277778	2.00	1.00
Hazen-Williams	591722	1.85	1.17
Scobey	610042	1.90	1.10

For the Darcy-Weisbach equation, K is given by the equation,

$$K = 0.811 \left(\frac{f}{g} \right)$$

where,

f = friction factor can be obtained from the Moody diagram

g = acceleration due to gravity (9.81 m s^{-2})

K for Hazen-William equation is computed by

$$K = (0.285c)^{-1.852}$$

C = Friction coefficient depends on pipe material and diameter.

K for Scobey equation is given by

$$K = \frac{K_S}{348}$$

where K_s = friction factor values depends on pipe diameter and pipe material.

There will be less friction loss along a pipe with several equally spaced discharging outlets such as submains and laterals than along a pipe of equal diameter, length, and material with constant discharge (constant discharge means that inflow to the pipe section equals the outflow from the section). This occurs because the quantity of water in the submain or lateral diminishes in the downstream direction because of outlet discharge (i.e. drippers or sprinklers attached with laterals).

The term F in equation ($h_l = FH_f + M_l$) equals 1 when there are no outlets between the up and downstream locations along a pipe (i.e. discharge along the pipe is constant). Equations $F = \frac{1}{m+1} + \frac{1}{2N} + \frac{\sqrt{m-1}}{6N^2}$ and $F = \frac{1}{(2N-1)} + \frac{2}{(2N-1)N^m} \sum_{i=1}^{N-1}(N-1)^m$ can be used to estimate F when there is more than one equally spaced outlet. Equation $F = \frac{1}{m+1} + \frac{1}{2N} + \frac{\sqrt{m-1}}{6N^2}$ is used when the distance from the pipe line to the first outlet is equal to the outlet spacing.

$$F = \frac{1}{m+1} + \frac{1}{2N} + \frac{\sqrt{m-1}}{6N^2}$$

Equation is used when the distance to first outlet is half of the outlet spacing.

$$F = \frac{1}{(2N-1)} + \frac{2}{(2N-1)N^m} \sum_{i=1}^{N-1}(N-1)^m$$

where,

m = Exponent, m depending on type of equation involved in estimating H_f.

N = Number of emitters

When the discharge varies widely from outlet to outlet, the equation is applied between successive outlets working from the known pressure to unknown pressure.

Design of Lateral, Sub Main and Main Pipes

Lateral Pipe Design

Drip irrigation lateral lines are the hydraulic link between the supply lines (main or submain lines) and the emitters. The emission devices can be connected directly to the lateral line (online or inline), mounted on a riser (micro sprinkler, jet) from a buried lateral or attached to the lateral on a tree loop. The lateral line will have hydraulic fittings (tees, unions, etc.,) to connect to the submain or main line. Lateral lines are usually made of LLDPE tubing ranging in diameter 12 mm to 16 mm. Laterals with only one diameter tubing are normally recommended to simplify installation and maintenance and provide better flushing characteristics. The procedures include, determining such lateral characteristics as: flow rate and inlet pressure; locating spacing of manifolds, which in effect sets the lateral lengths; and estimating the differences in pressure within the laterals.

In lateral line design a first consideration is acceptable uniformity of emitter flow or emitter flow variation. If the manufacturers variation is not considered, or assumed to be small, the design can be made to achieve a completely uniform emitter flow by using different emitter sizes or micro tube length. In general practice the emitter characteristics are usually fixed and the emitter flow rate is determined by pressure at the emitter in the line.

On fields where the average slope in the direction of the laterals is less than 3%, it is usually most economical to connect laterals to both sides of each manifolds. The manifold should be positioned so that starting from a common manifold connection, the minimum pressures along the pair of laterals are equal. Spacing of manifolds is a compromise between field geometry and lateral hydraulics.

Because of the possibility of laminar, turbulent or fully turbulent flow in drip laterals the Darcy-Weisbach equation should be used to compute head loss due to pipe friction. The Darcy-Weisbach friction factor, f, for small-diameter drip tubing is related to the Reynolds number, R_e the Reynolds number (R_e) is computed with the following equation.

$$R_e = \left(\frac{(\rho)(D)(V)}{(K)(\mu)}\right)$$

Where,

R_e = Reynolds number (dimensionless); D = diameter of pipe, cm;

ρ = density of water, g cm^{-3}; V = average velocity, cm s^{-1};

μ = viscosity of the fluid, N s m^{-2}; K = unit constant, 10 with these units

The equation used to compute friction factor (f) depends on the magnitude of N_R. For R_e less than 2000 (laminar flow), the friction factor:

$$f = \frac{64}{R_e}$$

For R_e between 2000 and 10,0000 (turbulent flow),

$$f = 0.32 R_e^{-0.25}$$

For R_e greater than 10,0000 (fully turbulent flow),

$$f = 0.80 + 2.0 \log\left(\frac{R_e}{\sqrt{f}}\right)$$

The Hazen-Williams equation with C =150 can also be used to estimate head loss due to pipe friction R_e > 1,00,000.

Submain Design

The submain line hydraulics are similar to that of lateral hydraulics. The submain line is designed to allow approximately the same energy loss as compared to the lateral line for several laterals and submain line. Keller and Karmeli (1975) recommended that the lateral energy loss should be 55 percent and the submain energy loss should be 45 percent of the total allowable energy loss.

The submain design depends on the location of flow or pressure regulation. Energy loss in the submain is directly related to the length of the submain line. The energy loss cannot exceed the allowable limits without lowering uniformity. On particularly steep slopes, each lateral may require individual pressure or flow regulation. In this case the length and diameter of the submain line are determined solely by balancing the energy cost and pipe cost. Since each lateral in this case is regulated, uniformity is independent of submain energy loss provided that submain losses do not interfere with the flow regulation.

The position of the inlet to the submain line depends on the field slope. Usually laterals are placed on contours, if possible with submains running with the prevailing field slope. With sloping submain lines, the inlet is positioned so that the uphill run is shorter than the downhill run. On gently sloping land or level areas the submain inlet should be located near the centre of the submain lines. Submain and main lines should be provided with either manual or automatic flushing valves. Each lateral connection at the submain should have a secondary filter screen to prevent entry of foreign material to the lateral and clogging the emitters.

The submain hydraulics characteristics can be computed by assuming the laterals are analogous to emitters on lateral lines. The hydraulic characteristics of submain and main line pipe are usually taken as hydraulically smooth since PVC pipe are normally used. The Hazen-Williams roughness coefficient (C) usually varies between 140 and 150. The energy loss in the submain can be computed with methods similar to those used for the lateral computations. The energy loss at the lateral connection will depend on the type of connection used, i.e., tee, elbow, bends etc. The total submain energy loss should include energy loss through filters, pressure valves, and other minor losses.

Mainline Design

Normally, flow or pressure control or adjustment values are provided at the submain inlet. Therefore, energy losses in the mainline should not affect system uniformity. The mainline pipe size is based on economic comparisons of power costs and pipe costs. The mainline pipe size should be selected to minimize the sum of power costs and capital costs over the life time of the pipeline.

Water Filtration for Irrigation Systems

Suspended solids need to be removed from within irrigation water to prevent clogging piping, valves, nozzles and emitters. These small particles may come from many sources. Surface and recirculated water may contain sand, soil, leaves, organic matter, algae and weeds. Ground water, although usually clean, may contain fine particles of sand or suspended iron.

Filter Selection

Before selecting a filter, a water analysis should be done to identify the type and quantity of solids. Keep in mind that seasonal changes such as algae growth or spring runoff can affect the loading. Chemicals and minerals in the water can also cause plugging. These may have to be addressed with chemical treatment methods.

The flow rate needed to supply the irrigation system should be then determined, taking into account the maximum water usage. Also consider the needs of any proposed expansion. Filters are available from 10 to more than 1000 gallons per minute capacity.

The level of filtration should next be determined (table). If you need water for an impact sprinkler system, having a 30-mesh screen filter may be adequate to remove leaves and trash. On the other hand, if the system you are supplying is a microirrigation system, a 200 mesh disk filter may be required.

The pressure loss created by the filter should be minimal. Pressure loss is related to the size of the filter opening and the water flow. Use the next filter size larger if the loss is excessive.

Screen Filters

These come in several sizes and shapes. Intake screens may be placed on the suction end of the pipe supplying water from a pond or stream to remove leaves or algae. Self-cleaning intake screens are available. These have internal high-pressure nozzles that clean the screen surface as it rotates.

In-line screen filters can be used as a final filtration if the water is fairly clean. They are low cost and available for a wide range of flows. The water passes through one or two cylindrical screen elements. Suspended particles are deposited on the outside. The most common models can be manually disassembled for cleaning. Models that are cleaned by turning on a set of brushes that wipe the screen are also available. Automatic hydraulic flushing of debris from the screen can be found on some models. Observe the difference in pressure between the inlet and outlet to determinine when to clean the screen.

Sieve bend screens are installed as a first stage to remove larger solids from return water from ebb and flow and hydroponic system water. The moving belt screen has also been developed for this purpose.

Disk Filters

The filtering element of a disk filter is made of a large number of flat, grooved rings that are stacked tightly together. The degree of filtration is determined by the number and size of the grooves. Intake water surrounds the filter element and is forced through the grooves, trapping the particles.

Cleaning is accomplished by reversing water flow. This expands the disk stack and a high-pressure water-air spray spins the disks throwing off the trash. This is dumped to a drain.

Disk filters are best suited for water sources with a low solids concentration. They have a low head loss and use only a small amount of water to backflush.

Table: Minimum filter opening to remove suspended solids.

Material	Mesh*	Opening (in.)
Leaves, twigs	30	0.023
Gravel	10	0.015
Course Sand	70	0.008
Fine Sand	600	0.001
Algae	2000	0.0002
Silt	3000	0.0001
*Human *hair is about* 750 mesh = 0,604 inches		

Media Filters

Media filters are best for removing organic matter, such as algae and slime, and fine inorganic material such as silt and clay. The filter is a steel or plastic tank containing sand, quartz or other inert material sized to provide the level of filtration desired. Incoming contaminated water flows through the bed depositing the solids. When the pressure loss reaches a predetermined level, backflushing begins. Clean water is forced up from the bottom causing a turbulent expansion of the media. Entrapped contaminates are loosened and are removed through a separate drain valve.

Centrifugal Separators

These filters are good for removing sand or other heavy matter from well water. They operate by introducing inlet water in a spinning motion inside a steel cone. Heavier particles as small as 200 mesh are forced by centrifugal action to the outside and slide down to a collection chamber at the bottom.

Centrifugal separators are low cost, create very little pressure loss and have high efficiency. They have no moving parts and can be arranged in parallel to increase capacity.

Surface Irrigation Hydraulics

Flow Regimes and Models

The process of surface irrigation combines the hydraulics of surface flow in the furrows or over the irrigated land with the infiltration of water into the soil profile.

The flow is unsteady and varies spatially. The flow at a given section in the irrigated field changes over time and depends upon the soil infiltration behaviour.

Performance necessarily depends on the combination of surface flow and soil infiltration characteristics.

The equations describing the hydraulics of surface irrigation are the continuity and momentum equation. These equations are known as the St.Venantequation.In general, the continuity equation expressing the conservation of mass, can be written as:

$$\frac{\partial Q}{\partial x} + \frac{\partial A}{\partial t} + I = 0$$

The momentum equation expressing the dynamic equilibrium of the flow process is:

$$\frac{1}{g}\frac{\partial V}{\partial t} + \frac{V}{g}\frac{\partial V}{\partial x} + \frac{\partial y}{\partial x} = S_0 - S_f + \frac{VI}{2gA}$$

Where,

      y - Depth of flow (m)

      t - Time from beginning of irrigation (sec)

      v - Velocity of flow as f (x, t) (m/s)

      x - Distance along the furrow length (m)

      I - Infiltrations rate as f (x, t) (m/s)

      g - Acceleration due to gravity (m/s²)

S_o- Longitudinal slope of furrow (m/m)

S_f- Slope of energy grade line (friction slope) in (m/m)

A - Cross-sectional area as f (x, t) (m²)

Q - The discharge (m³/s)

These equations are first-order nonlinear partial differential Eq. without a known closed-form solution. Appropriate conversion or approximations of these equations are required. So, several mathematical simulation models (Full hydrodynamic, zero-inertia, kinematic-wave and volume-balance) have been developed, however, among them volume balance models are more commonly used for design. The volume balance models consider only the continuity eq. ($\frac{\partial Q}{\partial x} + \frac{\partial A}{\partial t} + I = 0$) and ignore the momentum eq. ($\frac{1}{g}\frac{\partial V}{\partial t} + \frac{V}{g}\frac{\partial V}{\partial x} + \frac{\partial y}{\partial x} = S_0 - S_f + \frac{VI}{2gA}$).

Empirical Infiltration Equation

Infiltration rate affects surface flow as well as performance of irrigation. Several expressions have been proposed for expressing infiltration rate as a function of elapsed time.

The Lewis- Kostiakov Equation

A widely used empirical expression, for design of surface irrigation system, was originally proposed by Lewis (1937) but was erroneously attributed to Kostiakov.

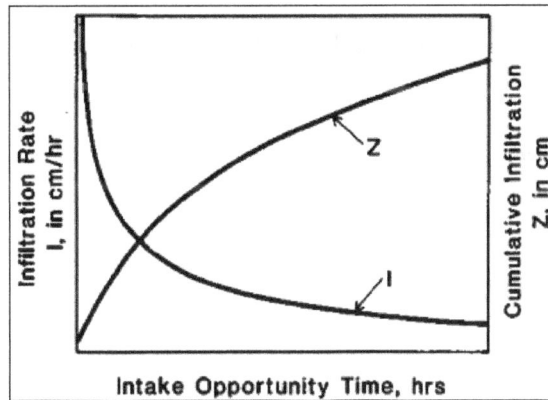

Infiltration rate and Cumulative infiltration vs. elapsed time.

Cumulative: $Z = kt^a$

$$Rate: I = \frac{dZ}{dt} = aKt^{a-1}$$

Where,

Z= the cumulative depth of infiltration or the volume of water per unit soil surface area

t = elapsed time

k and a = empirical parameters

i = the infiltration rate

Disadvantages of the original Lewis- Kostiakov Equation:

- It doesn't account for different initial soil water contents.

- For long infiltration times it erroneously predicts zero rate.

Infiltration rate never becomes zero instead it reaches a steady state or constant rate condition after a long time. Therefore, the above Eq. was modified to reflect the steady state infiltration rate which may occur during surface irrigation system with longer set times.

Modified Kostiakov Equation

The later problem can be fixed by adding a parameter representing a final infiltration rate (constant infiltration rate) to the previous Eq. ($\frac{\partial Q}{\partial x} + \frac{\partial A}{\partial t} + I = 0$) and ($\frac{1}{g}\frac{\partial V}{\partial t} + \frac{V}{g}\frac{\partial V}{\partial x} + \frac{\partial y}{\partial x} = S_0 - S_f + \frac{VI}{2gA}$). So the equation becomes:

Z = kt^a + f$_o$ t

$$I = \frac{dZ}{dt} = aKt^{a-1} + f_0 t$$

- Because of its simplicity, this model is frequently used in agricultural irrigation studies.

- Parameters k and a can be estimated by plotting the infiltration rate (I) or cumulative infiltration (Z) against time on log-log paper and fitting a straight line.

Volume Balance Concept

Volume balance equation

$$Q_0 t = \overline{A}x + \int_0^x Z(t - t_s)ds = \sigma_y A_0 x + \sigma_z Z_0 x$$

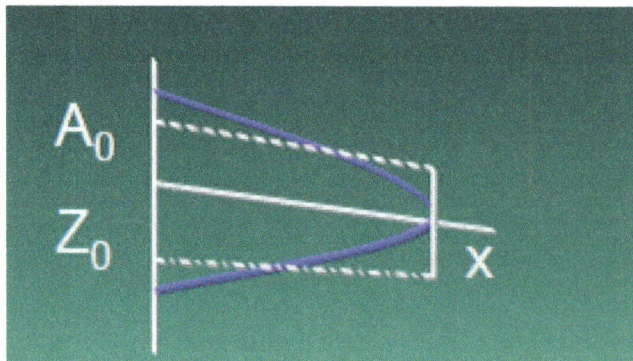

Approximation of sub-surface and surface profiles during volume-balance.

Also assume that advance characteristics follow a power function:

$$x = pt^r$$

Power Advance Volume Balance Model

Using power advance, Elliott and Walker (1982) gave the following solution to the volume balance considering Modified Kostiakoveq:

$$Q_0 t = \sigma_y A_0 x + \sigma_z x k t^a + \frac{t x f_0}{1+r}$$

Where, Q_0 = inflow rate, m³/min

A_0 = cross sectional area of flow inlet, m²

x = the advance distance, m

t= the advance time to distance x since beginning of irrigation, min

k and a= coefficients of modified Kostikov's Eq.

f_0= basic infiltration rate, m³/m/min

p and r =empirical parameters of advance curve

σ_y =surface storage factor and generally it has a value of 0.77

σ_z is defined as:

$$\sigma_z = \frac{a + r(1-a) + 1}{(1+a)(1+r)}$$

Advance Time Determination

Water will be distributed within a surface-irrigated field non-uniformly due to the differential time required for water to cover the field. To account for these differences in the design procedures, it is necessary to calculate the advance trajectory (curve).

It is first necessary to describe the flow cross section using two of the following functions:

$$A = a_1 y^{a2}$$

And

$$WP = b_1 y^{b2}$$

Or as a simpler substitute,

$$AR^{0.67} = p_1 A^p_2$$

Where,

A= cross sectional flow area, m²

R = hydraulic radius, m

y = flow depth, m

WP= wetted perimeter, m

a_1, a_2, b_1, b_2, p_1,p_2 = empirical shape coefficients

For border and basin systems,a_1, a_2, b_1 and p_1 are equal to 1. The value of b_2 is 0.0 and p_2 is 3.3333.

The next step is to determine the cross-sectional flow area at the field inlet. For sloping fields, this can be accomplished with the Manning Eq. as follows:

$$A_0 = \left(\frac{Q_0 n}{60 \rho_1 S_0^{0,5}} \right) 1/\rho_2$$

Where, Q_0 = Field inlet discharge, m³/min/unit width

n = Manning roughness coefficient

S_0 = field slope

The design input data required at this point are ,field length (L), S, nand. This information can be used to solve the volume balance Eq. for the time of advance,

$$Q_0 t_L - 0.77 A_0 L - \sigma_z k t_L^a L - \sigma_z' f_0 t_L L = 0$$

Where,

$$\sigma_z = \frac{a + r(1-a) + 1}{(1+a)(1+r)} \text{ and } \sigma_z' = \frac{1}{r}$$

Computation Steps of Advance Time:

1. The first step is to compute the flow cross-sectional area.

2. Make an initial estimate of power advance exponent (r) and label this value , usually setting , = 0.1 to 0.9 are good initial estimates. Then, a revised estimate of r is computed and compared below.

3. Calculate the subsurface shape factors.

4. Calculate the time of advance by Newton- Raphson technique.

    •   Assume an initial estimate of t_L as T_1, Then $T_1 = \frac{5(A_0 L)}{Q_0}$,

    •   Compute a revised estimate of t_L (say T_2) as,

$$T_2 = T_1 - \frac{Q_0 T_1 - 0.77 A_0 L - \sigma_z k T_1^a L - \dfrac{T_1 f_0 L}{1 + r_1}}{Q_0 - \dfrac{ak\sigma_z L}{T_1^{1-a}} - \dfrac{T_1 f_0 L}{1 + r_1}}$$

Compare the initial and revised of. If they are within about 0.5 minutes or less, the analysis proceeds to step 4. If they are not equal, let $T_1 = T_2$ and repeat steps b through c. It should be noted that if the inflow is insufficient to complete the advance phase in about 24 hours, the value of is too small or the value of L is too large and the design process should be restarted with revised values. This can be used to evaluate the feasibility of a flow value and to find the inflow.

5. Compute the time of advance to the field midpoint , using the same procedure as outlined in Step 4.

   • The half-length (0.5L) is substituted for 'L' and ' ' for ' '.

   • Volume balance Eq. is used with half length (0.5L) to find an appropriate value of

6. Compute a revised estimate of 'r' say,

$$r_2 = \frac{\log 2}{\log \dfrac{t_L}{t_0 .5\,L}}$$

7. Compare the initial estimate with the revised estimate:

   • If the difference between the two values is less than 0.0001 (error criterion), the procedure for finding is concluded.

   • If not then is replaced with and steps 3-6 are repeated until the prescribed error criterion is satisfied.

Intake Opportunity Time Determination

The basic mathematical model of infiltration utilized in the intake opportunity time determination is the Kostiakov- Lewis relation:

$$Z = kt^a + f_0 t$$

Where,

   Z- required infiltrated volume per unit length, m^3/m

   (per furrow or per unit width are implied)

   t- The design intake opportunity time, minute

   a -The constant exponent

   k - The constant coefficient, $m^3/min^a/m$ of length

   f_0 - the basic intake rate, $m^3/min/m$ of length

In order to express intake as a depth of application, Zmust be divided by the unit width. For furrows, the unit width is the furrow spacing, w, while for borders and basins it is 1.0. Values of k, a, f_0 and w along with the volume per unit length required to refill the root zone, Z_{req}, are design input data.

The volume balance design procedure requires that the intake opportunity time associated with be known. This time, represented by can be obtained from modified Kostiakov Equation using the Newton-Raphson procedure.

A step by step procedure for the calculation of intake opportunity time is given below:

1. Make an initial estimate of and level it

2. Compute a revised estimate of,

$$T_2 = T_1 - \frac{Z_{req} - kT_1^a - f_0 T_1}{-akT_1^{1-a} - f_0}$$

3. Compare the values of the initial and revised estimates of r_{req} (T_1 and T_2) by taking their absolute difference. If they are equal to each other or within an acceptable tolerance of about 0.5 minutes, the value of r_{req} is determined as the result. If they are not sufficiently equal in value, replace T_1 by T_2 and repeat steps 2 and 3.

Chapter 7

Irrigation System Design

The design of an irrigation system is aimed at minimizing damage and losses to soil, water, plant, air and animal resources. It also seeks to reduce deep drainage and runoff. Border check irrigation design and micro irrigation system design are a few examples of different irrigation systems designs. This chapter discusses in detail these theories and methodologies related to the design of irrigation systems.

A properly designed irrigation system addresses uniform irrigation application in a timely manner while minimizing losses and damage to soil, water, air, plant, and animal resources. The design of a conservation irrigation system matches soil and water characteristics with water application rates to assure that water is applied in the amount needed at the right time and at a rate at which the soil can absorb the water without runoff. Physical characteristics of the area to be irrigated must be considered in locating the lines and spacing the sprinklers or emitters, and in selecting the type of mechanized system. The location of the water supply, capacity, and the source of water will affect the size of the pipelines, irrigation system flow rates, and the size and type of pumping plant to be used. The power unit selected will be determined by the overall pumping requirements and the energy source available.

Key points in designing an irrigation system include:

- The irrigation system must be able to deliver and apply the amount of water needed to meet the crop-water requirement.

- Application rates must not exceed the maximum allowable infiltration rate for the soil type. Excess application rates will result in water loss, soil erosion, and possible surface sealing. As a result, there may be inadequate moisture in the root zone after irrigation, and the crop could be damaged.

- Flow rates must be known for proper design and management.

- Soil textures, available soil water holding capacity, and crop rooting depth must be known for planning and designing systemapplication rates, irrigation water management, and scheduling irrigations so that water applied is beneficially used by the crop.

- The water supply, capacity, and quality need to be determined and recorded.

- Climatic data - precipitation, wind velocity, temperature, and humidity must be addressed.

- Topography and field layout must be recorded.

- Farmer's preferences in irrigation methods, available operation time, farm labor, cultural practices, and management skills must be noted for selecting and planning the type and method of irrigation.

The most opportune time to discuss and review problems and revise management plans that affect design and operation of the irrigation system is during the planning and design phase. The physical layout of a system can be installed according to data from this guide. Operational adjustments then must be made for differing field and crop conditions.

Minimum requirements for the design, installation, and performance of irrigation systems should be in accordance with the standards of the Natural Resources Conservation Service, the American Society of Agricultural Engineers, and the National Irrigation Association.

Material and equipment used should conform to the standards of the American Society for Testing Materials (ASTM) and the Irrigation Association.

Irrigation System Design Criteria

When building, modifying, maintaining, or operating any part of an irrigation system, one must always consider the effect of that part on the entire system. The original design must be kept in mind and the balance of flows and pressures must be maintained.

Three overall design criteria should be kept in mind:

- Adequate availability of water: The water supply may be limited to some minimum flow rate for year round delivery. Wells may be fairly steady all year but surface water supplies may fluctuate. Determine the limits on the water supply and size primary and backup pumps to the available flow. Seasonal minimum flow rates will likely occur at time of maximum demand.

- Irrigation system capacity: An irrigation system frequently can not everything at one time and must be divided into zones that do not exceed the water supply available in gallons per minute. An irrigation zone is the amount of the irrigation system that can be operated at one time without exceeding the water available. The total number of zones can not exceed the daily water supply. An irrigation zone must be able to supply the water needed by a mature crop on a hot, summer day.

  An irrigation system should not be required to operate 24 hours a day to meet crop needs; some maintenance downtime should be included in the daily schedule. This extra capacity may be needed for extreme weather conditions.

- Distribution Uniformity: Uniformity refers to the extent to which all plants in an irrigation zone receive the same amount of water. Although perfect uniformity is an ideal that can not be achieved in practice, a high level of uniformity is desired so that the crop is fairly uniform.

Pressure losses in the piping system, emitter or sprinkler variability, and variability in plant water demand all contribute to the non-uniformity. Uniformity can be improved by investing in more expensive irrigation equipment, such as larger diameter pipes and more sprinklers to achieve better overlap of watering patterns. This investment may not be justified by the slightly better uniformity. However, the more uniform the irrigation system, the less over-irrigation is required.

Micro Irrigation Systems Design Considerations

Prior to designing a drip irrigation system, the following informations are needed to be assessed:

- A scaled plan of the site and area to be irrigated.
- Point-of-connection information, including static pressure and available flow.
- Irrigation water type (potable, non-potable, well, etc.) and characteristics.
- Soil type (important for determining drip line emitter and line spacing).
- Proposed planting, including relative water needs of all species, and sizes at planting and maturity.
- Local conditions, including elevation differences, local climate data (ETo), and other site specific information.

General Considerations

Several important design criteria affect drip irrigation system efficiency. The most important of these are:

- Efficiency of filtration.
- Permissible variations of pressure head.
- Base operating pressure to be used.
- Degree of control of flow or pressure.
- Relationship between discharge and pressure at the pump or hydrant supplying water to the system.
- Chemical treatment to dissolve or prevent mineral deposits.
- Use of secondary safety screening.
- Incorporation of flow monitoring.

Wetting Pattern

Drip irrigation systems normally wet only a portion of the horizontal, cross sectional area of soil. The percentage wetted area, P_w compared with the entire cropped area, depends on the volume and rate of discharge at each emission point, spacing of emission points and type of soil being irrigated. The area wetted at each emission point is usually quite small at the soil surface and expands somewhat with depth to form an inverted bulb-shaped cross section. P_w is determined from an estimate of the average area wetted at a depth of 150 to 300 mm beneath the emitters divided by the total cropped area served.

Systems having high P_w provide more stored water. For widely spaced crops, P_w should be held below 67% to keep the strips between rows relatively dry for cultural practices. Low P_w values reduce loss of water due to evaporation even where cover crops are used. Furthermore, it is costly to have a low P_w for more emitters and tubing are required to obtain larger coverage. However, closely spaced crops with rows and emitter laterals spaced less than 1.8m apart, P_w often approaches 100%.

The area wetted by each emitter, A_w, along a horizontal plane about 30 cm below the soil surface depends on the rate and volume of emitter discharge. It also depends on the texture, structure, slope and horizontal layering of the soil. The A_w values are given in the table for various soil textures, depths and degrees of stratification. They are based on daily or alternate irrigation that apply volumes of water sufficient to slightly exceed the crop water use rate.

On sloping land the wetted pattern may be distorted in the down slope direction. On the steep fields this distortion can be extreme; as much as 90% of the pattern may be on the down slope side. Spray emitters wet a larger surface area than the drip emitters. They are often used in the course textured, homogenous soils where wetting a sufficiently large area would require a large number of drip emitters.

Computing percentage wetted area:

For straight single-lateral systems, the percentage wetted area can be computed as:

$$P_w = \frac{N_p S_e w}{S_p S_r} \times 100$$

Where, P_w = Percentage of soil wetted, %

N_p = Number of emitters per tree

S_p x S_r = Plant spacing & row spacing, m

S_e = Spacing between emitters, m

w = wetted width, m

For spray emitters, the percentage wetted area can be computed as:

$$P_w = \frac{N_p \left[A_p + (S_e \times PS)/2 \right]}{S_p S_r} \times 100$$

Where, P_w = Percentage of soil wetted, %

N_p = Number of emitters per tree

S_p x S_r = Plant spacing & row spacing, m

S_e = Spacing between emitters, m

A_p = Soil surface area directly wetted by the sprayers, m²

PS = The perimeter of the area directly wetted by the sprayers, m

Irrigation Water Requirement

The irrigation water requirement for crop production is the amount of water, in addition to rainfall, that must be applied to meet a crop's evapotranspiration needs without significant reduction in yield. The crop water requirements under drip irrigation may be different from crop requirements under surface and sprinkler irrigation primarily because the land area wetted is reduced resulting in less evaporation from the soil surface. Most methods of estimating crop water requirement presently utilized provide estimates of evapotranspiration which probably contain a significant soil evaporation component.

Table: Expected maximum diameter of the wetted circle (Aw) formed by a single emission device discharging approximately 4 l/h on various soils.

Sand or root depth and soil texture	Homogeneous (cm)	Varying layers, generally low density (cm)	Varying layers, generally medium density (cm)
Depth 75cm			
Coarse	45	75	110
Medium	90	120	150
Fine	107	150	180
Depth 150cm			
Coarse	75	140	180
Medium	120	215	275
Fine	150	200	245

Estimation of Evapotranspiration

Weather parameters, crop characteristics, management and environmental aspects affect evaporation and transpiration. The principal weather parameters affecting evapotranspiration are radiation, air temperature, humidity and wind speed. Several procedures have been developed to assess the evaporation rate from these parameters. The evaporation power of the atmosphere is expressed by the reference crop evapotranspiration (ETo). The reference crop evapotranspiration represents the evapotranspiration from a standardized vegetated surface. The crop type, variety and development stage should be considered when assessing the evapotranspiration from crops grown in large, well-managed fields. Differences in resistance to transpiration, crop height, crop roughness, reflection, ground cover and crop rooting characteristics result in different ET levels in different types of crops under identical environmental conditions. Crop evapotranspiration under standard conditions (ETc) refers to the evaporating demand from crops that are grown in large fields under optimum soil water, excellent management and environmental conditions, and achieve full production under the given climatic conditions.

Factors such as soil salinity, poor land fertility and limited application of fertilizers, the presence of hard or impenetrable soil horizons, the absence of control of diseases and pests and poor soil management may limit the crop development and reduce the evapotranspiration. Other factors to be considered when assessing ET are ground cover, plant density and the soil water content.

Cultivation practices and the type of irrigation method can alter the microclimate, affect the crop characteristics or affect the wetting of the soil and crop surface. A windbreak reduces wind velocities and decreases the ET rate of the field directly beyond the barrier. The effect can be significant especially in windy, warm and dry conditions although evapotranspiration from the trees themselves may offset any reduction in the field.

The evapotranspiration rate from a reference surface, not short of water, is called the reference crop evapotranspiration or reference evapotranspiration and is denoted as ETo. The only factors affecting ETo are climatic parameters. Consequently, ETo is a climatic parameter and can be computed from weather data. ETo expresses the evaporating power of the atmosphere at a specific location and time of the year and does not consider the crop characteristics and soil factors. The FAO Penman-Monteith method is recommended as the sole method for determining ETo. The method has been selected because it closely approximates grass ETo at the location evaluated, is physically based, and explicitly incorporates both physiological and aerodynamic parameters.

The crop evapotranspiration under standard conditions, denoted as ETc, is the evapotranspiration from disease-free, well-fertilized crops, grown in large fields, under optimum soil water conditions, and achieving full production under the given climatic conditions.

The amount of water required to compensate the evapotranspiration loss from the cropped field is defined as crop water requirement. Although the values for crop evapotranspiration and crop water requirement are identical, crop water requirement refers to the amount of water that needs to be supplied, while crop evapotranspiration refers to the amount of water that is lost through evapotranspiration. The irrigation water requirement basically represents the difference between the crop water requirement and effective precipitation. The irrigation water requirement also includes additional water for leaching of salts and to compensate for non-uniformity of water application.

Crop Coefficient

Crop evapotranspiration can be calculated from climatic data and by integrating directly the crop resistance, albedo and air resistance factors in the Penman-Monteith approach. As there is still a considerable lack of information for different crops, the Penman-Monteith method is used for the estimation of the standard reference crop to determine its evapotranspiration rate, i.e., ET_o. Experimentally determined ratios of ET_c/ET_o, called crop coefficients (K_c), are used to relate ET_c to ET_o or $ET_c = K_c \, ET_o$.

The Kc coefficient incorporates crop characteristics and averaged effects of evaporation from the soil. Changes in vegetation and ground cover mean that the crop coefficient K_c varies during the growing period. The trends in K_c during the growing period are represented in the crop coefficient curve. Only three values for Kc are required to describe and construct the crop coefficient curve: those during the initial stage (K_c ini), the mid-season stage (K_c mid) and at the end of the late season stage (K_c end).

The amount of irrigation water requirement was estimated using the crop evapotranspiration (ET_c) which was calculated by the FAO Penman–Monteith method based on the climatic data. The FAO Penman–Monteith equation is as follows:

$$ET_c = K_c \frac{0.408\Delta(R_n - G) + \gamma\left(\dfrac{900}{T_{mean}} + 273\right)u_2(e_s - e_a)}{\Delta + \gamma(1 + 0.34 u_2)}$$

where ETc is crop evapotranspiration under standard condition (mm day^{-1}), Rn net radiation at the crop surface (MJ m^{-2} day^{-1}), G the soil heat flux density (MJ m^{-2} day^{-1}) which is relatively small and ignored for day period, T_{mean} the mean daily air temperature at 2 m height (°C), u2 the wind speed at 2 m height (m s^{-1}) (es – ea) the vapor pressure deficit (kPa), Δ the slope of vapor pressure curve (kPa °C^{-1}), γ the psychrometric constant (kPa °C^{-1}) and Kc the crop coefficient (varies between 0.45 and 1.05) which is affected by several factors such as crop type, crop height, albedo (reflectance) of the crop-soil surface, aerodynamic properties, leaf and stomata properties and crop stages.

Net Depth Per Irrigation

Normally, drip irrigation wets only part of the soil area. Therefore, the equations for determining the desirable depth or volume of application per irrigation cycle and the maximum irrigation interval must be adjusted accordingly. The maximum net depth per irrigation, d_x, is the depth of water that will replace the soil moisture deficit when it is equal to MAD. The d_x is computed as a depth over the whole crop area not just the wetted area; however, the percentage area wetted, P_w must be taken into account. Thus for drip irrigation equation can be given as:

$$d_x = \frac{MAD}{100} \frac{P_w}{100} W_a Z$$

Where,

d_x = maximum net depth of water to be applied per irrigation, mm

MAD = Management allowed deficit, %

W_a = available water holding capacity of the soil, mm/m

Z = Plant root depth, m

The net depth to be applied per irrigation, d_n, to meet consumptive use requirements can be computed by,

$$d_n = T_d f' \text{ and } f_x = \frac{dx}{T_d}$$

Where,

d_n = net depth of water to be applied per irrigation to meet consumptive use requirements, mm

f' = irrigation interval or frequency, days

f_x = average daily transpiration during peak-use period, mm

T_d = average daily transpiration during peak-use period, mm

For the design purposes, the T_d for the mature crop should be used for sizing the pipe network. Furthermore, assuming irrigation interval as one day, so that $d_n = T_d$, simplifies design process.

Gross Irrigation Requirements

Gross irrigation depth and volume requirements for drip systems are based on net requirements and efficiencies. The grass depth per irrigation, d, should include sufficient water to allow for unavoidable deep percolation. To minimize avoidable losses, systems should be well designed, accurately scheduled, and carefully maintained. Where $LR_i \leq 0.1$ or the unavoidable deep percolation is greater than the adjusted leaching water required $T_r \geq 0.9/(1.0 - LR_i)$.

$$d = \frac{d_n T_r}{EU/100} \quad or \quad d' = \frac{T_d T_r}{EU/100}$$

Where, $LR_i > 0.1$ or $T_r < 0.9/(1.0 - LR_i)$

$$d = \frac{100 d_n}{EU(1.0 - LR_i)} \quad or \quad d' = \frac{100 T_d}{EU(1.0 - LR_i)}$$

Where, d= gross depth of application per irrigation, mm

   d_n= net depth of water to be applied per irrigation to meet consumptive use requirements, mm

   d' = maximum gross daily irrigation requirement, mm

   T_r = peak use period transmission ratio

   T_d = average daily transpiration during peak-use period, mm Eu = emission uniformity, %

   LR_i = leaching requirement under drip irrigation

The gross volume of water required per plant per day, G is a useful design parameter for selecting emitter discharge rates:

   G= K d' S_p S_r

where,

   G = gross volume of water required per plant or unit length of row per day, L/day

   K = Conversion constant, which is 1.0

   d' = maximum gross daily irrigation requirement, mm

   S_p = spacing between plants, m

   S_r = spacing between row, m.

Capacity of Drip Irrigation System

It is necessary to determine the system capacity and operating time per season to design a pumping plant and pipeline network that are economical and efficient. The capacity of the drip irrigation system, Q_s is the maximum number of emitters operating at any given time multiplied by average emitter discharge, q_a. According to Keller and Bliesner for uniformly spaced laterals that supply water uniformly spaced emitters:

$$Q_S = K \frac{A}{N_S} \frac{q_s}{S_r S_l}$$

Where,

Qs= Total system capacity, Ls⁻¹ q_a=Average emitter discharge, L/hr

K= Conversion constant, 2.778 S_e = Emitter spacing, m

A= Field area, ha S_l = Lateral spacing, m

N_s = Number of operating stations

Some systems require extra capacity because of anticipated slow changes in q_a can result from such things as slow clogging due to sedimentation in long path emitters or compression of resilient parts in compensating emitters. Both decrease and increase in q_a necessitates periodic cleaning or replacement of emitters. To prevent the need for frequent cleaning or replenishment of emitters, where decreasing discharge rates are a potential problem, the system should be designed with 10 to 20 % extra capacity.

Border-check Irrigation Design

Border-check irrigation of perennial pasture can be quite efficient – on suitable soils, with an appropriate layout and good management.

Efficient irrigation is applying the water needed by the pasture with a minimum of deep drainage or runoff. Irrigation efficiency can be quantified by the amount of the applied water actually used by the pasture (the application efficiency), and the "evenness" of the application (the distribution uniformity).

The design of an efficient border-check irrigation layout depends on many interrelated factors, including:

- The soil moisture deficit at the start of the irrigation;
- The soil infiltration rate, which is partly dependent on the soil moisture deficit;
- The slope of the bay;
- The length and width of the bay, and hence the area of the bay;
- The hydraulic roughness of the bay surface;
- The flow rate applied; and
- The time that it is applied for - the time of cut-off.

Soil Moisture Deficit

A border-check irrigation system must be well designed and managed to be efficient.

This is the depth of water needed to "refill" the pasture rootzone to "full" (field capacity). The recommended best management practice for irrigating perennial pasture by border-check irrigation in the Shepparton Irrigation Region (SIR) is to irrigate after 50 mm of pan evaporation less rainfall (E-R) has occurred since the previous irrigation. This is equivalent to about 40 mm of pasture water use. This is the target irrigation application, or the depth of irrigation needed to refill the rootzone.

A higher soil moisture deficit (a greater E-R interval; resulting in a drier soil profile) will increase the depth of water taken up by the soil during irrigation. However, higher moisture stress in the pasture will reduce its water use and productivity.

Infiltration Rates

For most SIR soils, infiltration is typically quite rapid initially, before stabilising at a relatively low constant rate.

The initial rapid wetting up of the soil is known as the crack-fill part of infiltration, and is largely dependent on the soil moisture deficit. Typically, crack-fill is about ¾ of the soil moisture deficit, which for the above 40 mm soil moisture deficit is about 30 mm.

The ongoing final infiltration rate typically ranges from less than 1 mm/hr for heavy clays to 5 mm/hr for fine sandy loams, and is independent of the soil moisture deficit.

Considerable variability occurs with crack fill and final infiltration rate components, both between and within soil types, and within paddocks.

The infiltration characteristics of a given site can also change with time, subject to management. For example, a site that has not been irrigated for some years can develop cracks in the sub-soil which allow higher than expected infiltration rates, but these can slowly decrease with irrigation over a season or even longer as the sub-soil wets up and swells. Shallow watertables (say 0.5 to 1.0 m below the surface) can restrict infiltration, particularly in soils that would otherwise have relatively high final infiltration rates.

A study of soil hydraulic properties in the SIR (Mehta and Wang, 2004) measured the final infiltration rate of the Bhorizon subsoil, which determines the final infiltration rate of the soil.

Ideally, bays should contain only one soil type, or at least, soils types that have similar infiltration characteristics.

Soils with high final infiltration rates are generally not suited to border check irrigation.

Slope

This is largely determined by the site's topography, but can be altered by earthmoving.

Slope is important for drainage of excess water, particularly on medium to heavy soils. However, bays that are too steep can be prone to erosion and difficult to cover with water.

Slope affects the rate of the irrigation moderately, but has less impact on irrigation performance than the effects of infiltration rate, flow rate and bay length.

Bay Length

Bay length is often determined by the topography, supply channel and drain infrastructure, or property boundaries. A minimum bay length of 300 m is generally recommended to facilitate farm management, although shorter bays can be efficiently irrigated and may be appropriate in particular situations.

The maximum bay length recommended depends on the final infiltration rate.

For most SIR soils with relatively low final infiltration rates, surface drainage following irrigation or rainfall is the major constraint to bay length. With higher infiltration rate soils, excessive infiltration and poor distribution uniformity are more important considerations.

Bay Length and Flow Rates

Guide to the optimum bay flow rates for typical bay lengths and infiltration categories. The values have been derived using the Analytical Irrigation Model (AIM) developed by Austin and Prendergast.

Assumptions:

- Target application of 40 mm, crackfill of 30 mm.

- Surface roughness of 0.3.

- Slope of 1:700, minimal runoff (1% to 3%).

Flows of less than 0.1 ML/d/m are generally not recommended because the shallow depth of flow makes full coverage of the bay difficult with even a moderate slope and good grading.

Flow Rate and Bay Width

Normally, the design flow rate adopted is the highest normally available from the water supply, to maximise irrigation labour efficiency. Ideally, bays are designed to take the whole supply flow to maximise labour efficiency, minimise the number of farm channel structures and facilitate automation. Where the flow rate available exceeds that required for the selected bay width, two or more bays may be irrigated together.

Bay Width

The total bay width needed to achieve specified flow rates per metre width of bay with various supply flow rates. However, there are practical constraints on bay width and area:

- The minimum bay width is determined by the equipment used to construct the bay. Typically, a laser grader requires at least 30 m width to operate efficiently, and this is generally recommended as the minimum bay width. However, 20 m may be practical with smaller equipment.

- The maximum bay width is limited by the desirability of achieving full coverage of the bay from one bay outlet, and economically by the high cost of earthmoving likely to be needed to achieve very wide bays. The coverage from the bay outlet depends on the flow rate, the slope, and the depth of flow, which depends on the surface roughness.

- The bay area (length x width) is ideally the required rotational grazing area or a multiple of it.

Surface Roughness

The rate that water moves down the bay and the depth of flow on the bay depend partly on the density of the crop being irrigated; e.g. water moves faster and shallower through a stalky wheat crop than through a leafy dense pasture. This is not normally an issue considered by irrigation designers, but is relevant where an irrigation model (such as AIM) is used.

Surface roughness is expressed as "Manning's n"; a roughness coefficient used in hydraulic design. For perennial pasture, Manning's n values of 0.2 to 0.4 are common.

Application Time:

This is the time interval that water is applied to the bay for, or the cut-off time. It is the time required to apply the volume of water needed at the design flow rate.

Application times of *2 - 6 hours* are common. From table, *4 hours* is a desirable maximum (for 500m long bays on low infiltration rate soils). Shorter bays and higher final infiltration rate soils require shorter application times.

While some runoff is desirable to ensure that the whole bay is irrigated uniformly, too long an application time results in excessive runoff. While runoff is not wasted where it is collected in a drainage reuse system, excessive runoff (greater than say 5 – 10 % of the target application) is undesirable, as water is on the bay surface for longer than necessary, potentially resulting in excessive infiltration or waterlogging.

The time to cut off the flow onto the bay is normally judged from experience, perhaps fine tuned by knowledge of the soil moisture deficit, and by the observed rate at which water advances down the bay. Typically, the optimum cut-off time is when water has advanced to half or two-thirds the length of the bay.

The intake opportunity time is the time that free water is on the surface of the bay. It is longer than the application time, and varies along the bay. A good border-check irrigation design results in the opportunity time being relatively uniform along the bay and just long enough to allow the required depth of water to infiltrate. This results in a relatively uniform irrigation with little deep seepage.

Runoff and Drainage Reuse

Collection and storage of runoff in a reuse system is essential for efficient irrigation. It is also critical for the effective management of nutrients to prevent them from leaving the property.

While a border-check irrigation system can be potentially very efficient, it must be managed appropriately to achieve that efficiency and to achieve its potential productivity.

Design of Sprinkler Irrigation System

Step 1: Gather Information

Decide whether you want a manual or auto-matic system. The process would be the same, except that a controller, automatic valves and wire are necessary to operate an automati-cally control-led system.

We recommend if it is a large or complex system, you should get the help of a Certified Irrigation Designer or a qualified landscape architect.

Find out the following information:

a. Static water pressure _____ psi

If you have city water call your local water department or use pressure gauge on garden valve output nearest area to be watered—no water should be running in or around the house.

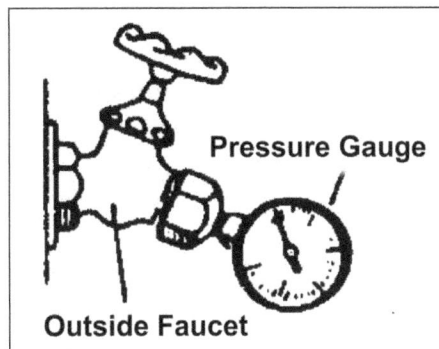

b. Water meter size

Check meter or call local water department.

c. Gallons per minute: Use chart below if you have city water or if you have a pump use the following: Run hose bib until pump turns on, then measure gallons per minute (the time it takes to fill a 5 gal. or other known quantity bucket) and note pressure on pressure gauge on the side of the pressure tank. Use the following formula to determine gallons per minute (GPM): divide the container size (in gallons) by the seconds it takes to fill the container and multiply by 60. Example: if it takes 30 seconds to fill a 5 gal. bucket, we have 10 GPM available.

d. Plumbing regulations, building permits needed. Call city hall or your local building department.

e. Inside diameter of pipe (service line) running from water meter or pump to house. Wrap a piece of string around the outside of the service line. Measure the length of string required to encircle it. Look up service line size from following chart.

f. Determine backflow prevention device requirements of city or water system. A cross connection creates the possibility of back siphonage or contamination of potable water, and a backflow preventer is a must. High hazard conditions exist where a fertilizer injector is installed.

Step 2: Plan Your System

Use a Planning Grid to Make an Accurate Scale Layout of Your Property:

a. Include location of house, driveways, walkways, paths, fences, walls, structures, planters, patios, flower beds, shrubbery and lawn areas.

b. Shade in all area to be covered by water: lawns, shrubbery, special grounds, etc.

c. Indicate water supply locations: spigots, well, water service entrance, etc.

Step 3: Select and Position Sprinkler Heads

a. It identifies the major types of standard heads, gives their uses and gallons per minute (GPM), tells what patterns are available, and suggests correct spacing requirements.

Step 4: Determine Size, Number, Type and Location of Valves

a. Begin graphing large lawn areas: Place quarter circle heads in the corners. Place half circle heads between corners, following recommended equal distant spacings. Place full circle heads within each perimeter, the number will depend on the perimeter's overall dimensions. Study Diagram below, a typical yard area showing positioned sprinklers. Impulse pop-up spinklers may be used if area size permits.

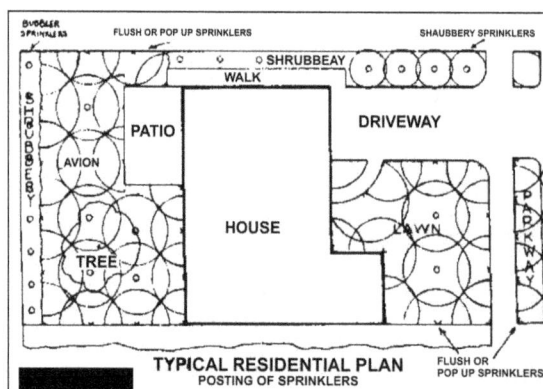

TYPICAL RESIDENTIAL PLAN
POSTING OF SPRINKLERS

b. Now position heads in small lawn areas, parkways, and at the side of your house. These areas are usually watered by one or two rows of part circle heads.

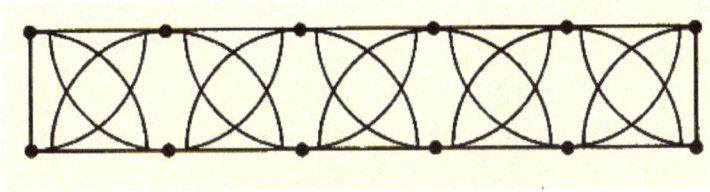

c. Complete graphing of heads by positioning them in flower beds, shrubbery areas, planters and other special places.. Shrub head patterns are equivalent to regular head patterns. For good coverage you want to have 100% overlap of heads.

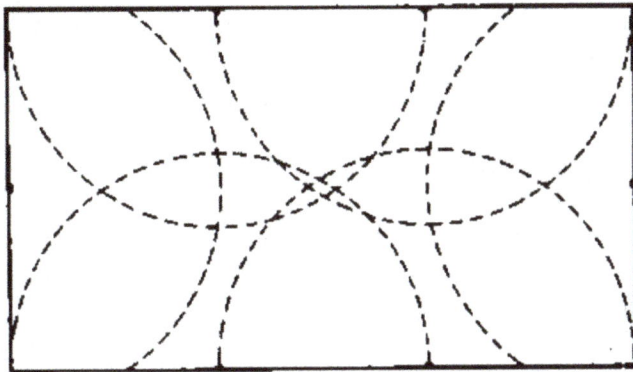

d. Irrigation heads and bubblers should be used in tree wells, planters, for ground cover, and around plants that require soaking.

- Determine the sizes, number, type, and location of control valves:

Valve sizes are determined by the size of your water meter and the size of your service line (refer to Step One). Your valve sizes do not need to exceed your service line size.

The type of valve depends on local plumbing regulations. Most cities require an anti-siphon control valve to prevent dirt and foreign matter from draining back into the main water system.

Valve locations are determined largely by their nearness to your service line and existing outlets. Pick a convenient location away from heavy foot traffic. Group your valves together to save on supply line pipe and other materials.

Step 5: Determine Size of Supply Line

a. The supply line is the piping from your water source such as your water meter or well, the line servicing the dwelling, or the existing outlet shown in Diagram 10- to the sprinkler valves.

b. Because of friction loss in the supply line, it is necessary to consider the distance from the connection to the valve location. If the tie-in point to your water meter, service line, or existing outlet is within 50 feet of the valve locations, use a supply line pipe equal to the size of your largest control valve. If between 50 and 100 feet, use one size larger than your largest control valve. If an existing outlet is being used, it must be as large as your control valve. The situation may arise where your supply line is larger than your control valve. In this case, it is necessary to reduce the supply line reducers to the size of the valve.

c. Determine the type of pipe you wish to use: It is recommended that PVC Schedule 40 be used to supply valves and then PVC Class 200 or poly-pipe after the valves, for residential systems. PVC is semi-rigid and comes in 20' lengths with glue type joints. Poly-pipe is a flexible hose type pipe and comes in 100' or larger coils. This pipe is attached with insert fittings rather than glued joints. Poly-pipe is somewhat less susceptible to freeze damage. Warning: do not use poly-pipe as the connecting pipe between the service line and control valves. Surge pressure may rupture poly-pipe.

d. Once the pipe is selected, draw lines on the plot plan connecting the sprinklers in each zone to eachzone valve. The gallons per minute (GPM) used by any zone may not exceed the total capacity of your home's water system. If it does, the circuit will not operate properly. Use the following chart to size the pipe lines you have drawn in, using the total GPM for each zone.

Step 6: Group Sprinklers and Graph Plan

a. A sprinkler group is the number of sprinkler heads adjacent to each other that will work off of the same control valve. If at all possible, all sprinklers in a group should be of the same function. For example, only bubblers should be grouped together. Start at the end of each particular sprinkler group and work towards its control valve, noting the number of planned heads. If you have three valves you will have three groups of sprinklers.

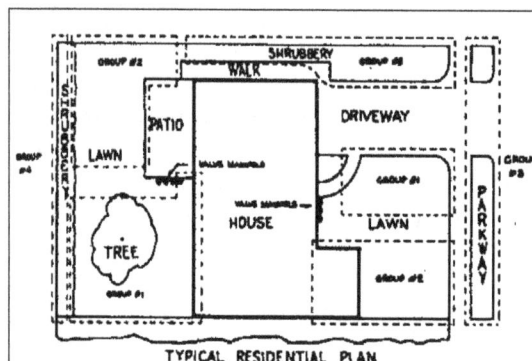

TYPICAL RESIDENTIAL PLAN

b. Determine the number of circuits needed: Maximum Flow chart above shows the maximum GPM flow that may be drawn by a circuit and still assure proper operation. This flow rate depends on 3 factors: the service line size, the static water pressure, and the type of sprinkler used. Refer to the information you gathered earlier and use the Maximum Flow chart to determine the GPM available for the operation of one circuit.

c. Start at the control valve and layout the piping system: Draw a feed line through the middle of each sprinkler group. Draw in side branches to adjacent heads. Diagram shown for a typical piping layout.

Example: Determining Gallons/Minute chart indicates that with a 3/4" supply line and 40 psi, you have 9 gpm available.

Step 7: Set The Controls

Product Guide indicates that each full circle (360°) spray head uses 3 gpm. If you divide the 3 gpm into the 9 gpm (9 divided by 3=3), or 3 full circle spray heads on each valve. If you use all half circle sprays, you use 6 sprinklers (9 divided by 1.5 = 6).

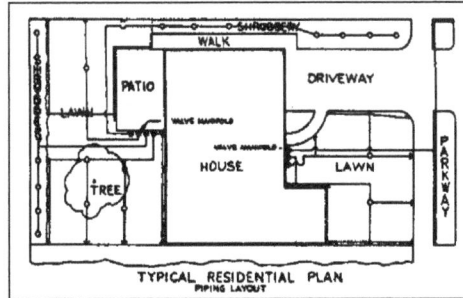

Lay out the piping: In your system piping will run from the service line (source of water) to the first set of valves, then from the first set of valves to the second set, etc., and from the valves to the sprinkler heads.

Draw in these connecting pipes on your grid layout and follow these rules:

a. Use as many straight runs as possible.

b. Avoid turns which result in friction and loss of pressure.

c. Avoid going under sidewalks and driveways whenever possible.

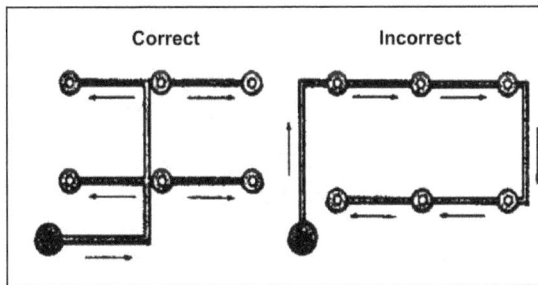

To avoid forcing water through too many turns (right), design connecting pipe so that several lines branch from the first head in the circuit (left).

Review your plan, thenorder what you need:

a. Check over the details of your plan carefully, then order selected sprinkler heads and valves.

Step 8: Set your Control Valves

a. To connect the supply line, shut off water at the meter, cut a section of the service line and insert a standard slip coupling tee. Run the supply line to the location of your valves. To connect to an existing garden hose outlet, simply remove the valve and insert a nipple and a tee. Replace the hose valve and run pipe down to the sprinkler valves.

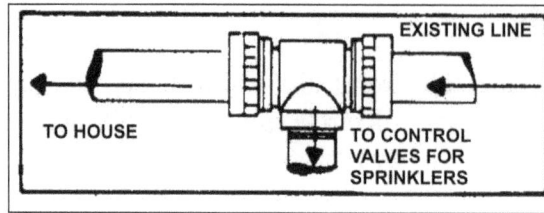

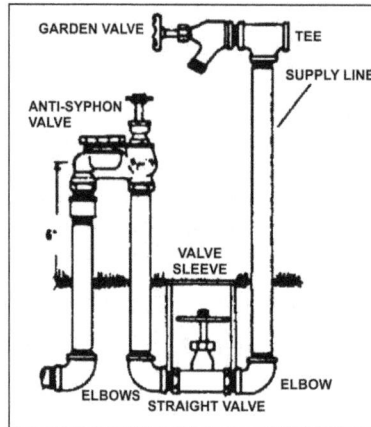

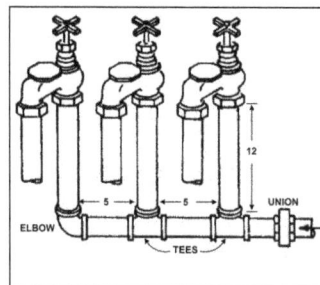

b. A group of valves is called a manifold: Diagram shows an example of a manifold for three valves. In a manifold, use fittings and pipe the same size as your valves. If more valves are needed, additional fittings would be added. Always install a master shutoff valve before your control valve manifold.

c. Install your valves, making certain that anti-siphon valves are 6" above the ground and 5" apart to allow for an adapter to automate the system at a later date. Angle valves should be buried. Refer again to Diagram for typical valve installations for a single valve or refer to manifold valve hookup.

d. Do not install anti-siphon valves under constant pressure.

Step 9: Position Sprinkler Heads

a. Using stakes (or sprinklers) and string (for piping), mark the locations of your heads and piping. Important-Maximum GPM flow may vary significantly because of pressure losses resulting from size and type of pipe used as well as the total length of pipe in an installation. Another important factor is elevation, because you will lose approximately 1/2 psi per foot of elevation going up a hill, and gain approximately 1/2 psi per foot of elevation going down a hill. Therefore, before backfilling any trenches, hook up each circuit and test for sufficent coverage.

b. Dig trenches 5" to 6" deep, putting any sod on one side and any dirt or soil on the other side of the trench. For cold climates, lay the pipe so the water will flow to a drain valve which should be installed at the lowest point. Hard ground should be watered two days before trenching.

Trenching machines are an easier, faster alternative to digging with a spade. They can be rented by the hour, day or week-usually from a lawn-supply store or rental equipment dealer. The person you rent from can show you how to operate the machine properly and safely. Trenchers should not be used to dig through ground cover, flower beds, on steep slopes or near buildings.

Going under obstacles—a hose may be used to tunnel under brick and concrete walks. Attach a piece of galvanized pipe to your hose with a hose-to-pipe adapter. Point the end of the pipe where you want to tunnel. Then turn on the water. Push the pipe under the concrete; the force of the water will blast away soil to form a tunnel. Tunneling requires care to avoid damage to walks and driveways.

c. Locate valves: Valves can be located above ground or, to keep them out of sight, below ground. Use stakes to mark the locations of the valves as indicated on your grid diagram.

Protect valves by sheltering them in boxes. You can buy boxes or build them yourself. If you buy boxes,check to see how many valves each box holds. If you build the boxes, we recommend using high-quality redwood because it resists rot.

d. After hooking up, close all the control valves on your manifold and open the master valve. Check all connections for leaks.

Step 10: Connect Piping

a. After all trenching is finished, connect all threaded fittings first, starting at the water source. Then connect all slip or insert fittings. If using galvanized pipe, use an adequate amount of pipe joint compound; if using plastic pipe, be certain of a leak-proof seal.

b. How to cut and connect PVC pipe:

1. Cut PVC pipe to desired length with a hacksaw or PVC cutter. Wipe cut clean and remove burrs.

2. Apply solvent and glue to both pipe and fitting.

3. Push pipe as far as possible into fittings. Make 1/2 turn for sure sealing.

4. Wipe off excess solvent and glue.

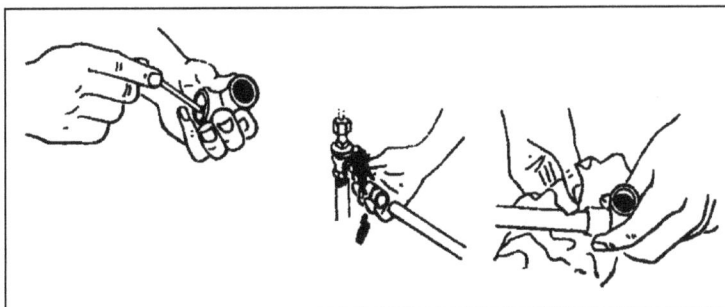

c. Locate valves: Valves can be located above ground or, to keep them out of sight, below ground. Use stakes to mark the locations of the valves as indicated on your grid diagram.

Protect valves by sheltering them in boxes. You can buy boxes or build them yourself. If you buy boxes,check to see how many valves each box holds. If you build the boxes, we recommend using high-quality redwood because it resists rot.

d. After hooking up, close all the control valves on your manifold and open the master valve. Check all connections for leaks.

Step 11: Flush System

a. Flush the main line: Wait for the solvent to dry (about one hour). Close all valves except the one at the end of the line. Turn on the water. Flush until the water runs clear. Shut off the last valve and check for leaks.

b. Flush the valves: This removes any dirt or soil that may have gotten trapped in the valves during installation. Open the valves using manual bleed finger screws or knobs. Turn on the water and flush valves until the water runs clear. Check for leaks and proper operation.

c. Install the automatic system timer: Locate the timer in your garage or some other convenient place. Make sure an adequate power supply is available. Timers require only a standard AC outlet.

Run wires along the trench from the valves to the system timer. Take one wire from each valve and connect to a common terminal on the timer. Take the other wire from each valve and connect one wire per terminal to the other terminals in sequence. See the instruction sheet that comes with your system timer for full details. Plug in the timer. Test the system by electronically opening and closing each valve in sequence. When the controller has been installed and fully tested, fill in the trench with dirt; then replace the sod.

d. Flush the system: Seal risers with pipe plugs. Leave the riser at the end of the line unplugged. Turn on the water, open the valves, and flush until the water runs clear. Remove the pipe plugs.

e. Install sprinkler heads: Attach sprinkler heads to risers. Check your grid diagram to make sure you attach the right head to each riser. Check the height of the heads. If necessary, cut risers to adjust head height. The top of the sprinkler should be flush with ground level. Re-install the sprinkler heads.

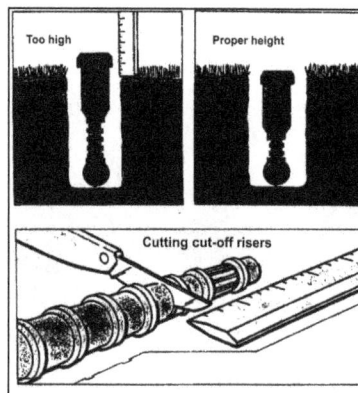

Step 12: Check System Operation

Turn on the water and open the control valve. If coverage is incomplete:

1. Make sure the control valve, main valve and shut-off valve are open all the way.

2. Turn off any water being used in the house (washers, showers, sinks).

3. Adjust the screws on the sprinkler heads to fine-tune the spray pattern.

4. If coverage is still not complete, go back and check your system layout against the plans.

5. When coverage is satisfactory, fill in the trench with dirt and cover with sod.

6. You've just installed a working circuit. To finish the job, repeat these steps for all circuits in the system. Relax and install the system one circuit at a time.

Hints on Operating your System

1. Operate only one valve at a time.

2. Water when pressure will be the greatest. This will probably be early in the morning.

3. Water shrubs, trees and plants deep and less frequently.

4. Water lawn areas for shorter periods and more frequently.

5. Do not water in the heat of the day to cut down on water evaporation. This will also eliminate the burning of plant and lawn areas.

6. Sprinklers should be cleaned periodically to insure proper functioning.

Design of Drip System

Drip irrigation is the most efficient method of water supply, achieving over 90% efficiency. With drip irrigation, water is applied directly to the crop root zone, where soil soaks it immediately, thus reducing evaporation or runoff.

Efficient irrigation of strawberries with dripline system.

Despite its efficiency, prior to drip system installation, a farmer must be aware of the following limiting factors:

1. Soil type (clay, loam or sand); important for determining dripline emitter and line spacing.

2. Planting scheme— dense, sparse or mixed plantings.

3. Topography (sloped or flat); used to determine inline tubing orientation and modify spacing or separate hydro-zones where needed.

4. Irrigation water type and characteristics.

5. Proposed planting, including relative water needs of all species, and sizes at planting and maturity.

6. Local conditions, including elevation differences, local climate data (ETo), and additional site specific information.

All of the aforementioned factors contribute to specific requirements regarding specific drip irrigation designs and components.

Drip Irrigation System Components and their Specific Function

The drip irrigation system is a set of various components which, combined together, provide water to plants in the most efficient way. This irrigation type suits any landscape and crop type, whether tree, vegetable, arable crop, or grass. It's mostly used in farmland with runoff problems. Beneficial, it can be installed both above and below the soil surface.

Irrigation system in young vineryard installed 50 cm (19.7 in) above the ground.

Typical drip irrigation system consists of the following components:

Pump Unit

The pump unit takes water from the source and provides the appropriate pressure for delivery into the pipe system.

Control Head

The control head consists of valves and filters. Some control head units also contain a fertilizer or nutrient tank for the slow adding of measured fertilizer doses into the water during irrigation.

Valves control the discharge and pressure in the entire system. There are two different kinds of valves which can be installed to control the water flow:

- Isolation Valves –used for infrequent shut off of water supply, i.e. to shut off water for repairs or during non-irrigation season. Isolation valves are usually manually operated.

- Control Valves – used to irrigate individual areas that are separate from one another. There can be several control valves installed on drip irrigation.Control valves can be automatic or manual. Automatic are wired to a controller or solar powered.

A backflow preventer is a device which prevents the irrigation system water from being siphoned back into drinking water. To ensure a constant level of lower water pressure, the drip irrigation system needs to have a pressure regulator installed. It reduces water pressure up to 60 psi, in order to ensure regular operation of all drip components.

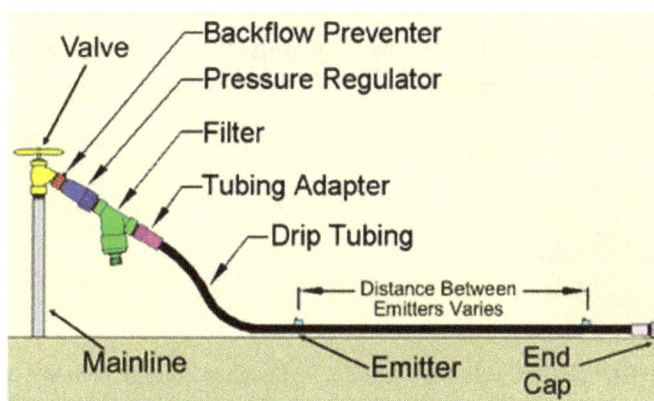

Main components of drip irrigation system.

Filters are the head of the drip irrigation system. They clear the water and keep dirt and debris from clogging the drip emitters. Common types filters include screen filters and graded sand filters, which remove fine material suspended in the water.

Main, Submain Lines and Laterals

The main lines are the pipes which supply water from its source to valves. They are made of galvanized steel, copper, polyethylene (PEX), or PVC. Since PVC can be easily damaged by solar radiation, these main lines should be buried in the soil.

Submains and laterals supply water from the valve to the drip tube. They are usually made of PVC and polyethylene and therefore should be buried below the ground to prevent sun damage. Lateral pipes are usually installed on large drip systems, where multiple drip tubes are needed. Additionally, in small systems, the drip tube is connected directly to the valve.

Drip tubing is the special polyethylene or rigid PVC tube which transfers water to plants. It's placed on the ground surface between the plants. Along its length, it has placed emitters which water the plants. Drip tubing systems are designed to last for 10 to 20 years or more. For this reason and the associated high initial cost, they are mostly used for permanent crop installations (fruit trees and vines).

Drip tubing in which emitters can be installed.   Drip tubing with already installed emitters.

Drip tape lasts for only a short time, from one season to a few years. Therefore, it's most frequently used for row and field crops. Drip tape can be placed on the soil surface or with just below, enough so that the soil is protected against the wind and sun.

Both drip tubing (with in-line or built-in emitters) and drip tape can be installed for subsurface irrigation as well.

Micro, or spaghetti tubing, is an additional small tubing which delivers water directly to individual plants, from the drip tubing to the emitter.

Drip tape with already created holes for water application.   Spaghetti tubing installed on drip irrigation system.

All of these parts are connected to the main drip pipe with a swivel (tubing) adapter.

Emitters or Drippers

Emitters are the main component of a drip system. These small devices control the water flow to the plants. They can be screwed or snapped onto the drip tube. Common practice is to apply 1 or 2 emitters per plant, depending on the crop size and row space. Hence, fruit crops usually have 1-2 emitters installed per plant, widely spaced, while in row crops, emitters are more closely spaced in order to wet a strip of soil.

Emitters can be installed on a drip tube by creating a hole in the drip tubing using a punch. Care should be taken to use a special punch designed for a certain emitter so as to avoid creation of big holes and water leaching. Some emitters are self-piercing, meaning they do not require the use of a punch. There are also drip tubes with already installed, uniformly spaced emitters inside the tube.

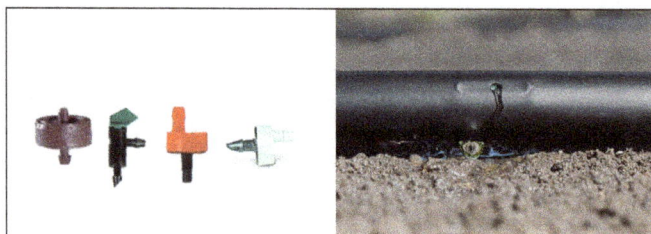

Various types of drip emitters.   Dripline with already installed emitters.

There are various emitter types which differ in flow rate, pressure, method of installation of drip tube, and design. The main benefit of every emitter is to provide a specified and constant water discharge that will not vary too much with pressure changes and won't be blocked easily.

Additional Drip System Components

Tee Fitting - allows splitting drip line into separate lines flowing to different locations.

End cap – end of the main drip line in an irrigation system. The end cap is simply placed on the end of the line (tubing), then that line is bent and fed through the end cap, "pinching" off the main line.

Coupler Fitting – allows the connection of two pieces of tubing in order to continue a drip line to the desired location.

Air vent - prevents air from being sucked into the emitters when the system is turned off. It should be installed at the highest point on the drip tube to prevent it from being covered with dirt. The air vent is usually installed on large drip systems which are on a slope, as the elevation change creates a more powerful suction that will suck in more dirt.

Fertilizing system – used to add fertilizers and pesticides when watering the plants. This farm practice is called fertigation.

Drip irrigation installed for vegetable crop production.

Due to the localized water application, drip irrigation system is a great solution for smart irrigation water management. It benefits both the farmer and a crop. Drip irrigation enables a farmer to achieve an easy and inexpensive installation, reduce weed growth, reduce insect pest and disease occurrence, and facilitate proper water management. Moreover, it can be used in all climates and on every soil and crop type.

References

- System Design Criteria: cluster47.canvas-user-content.com, Retrieved 8 June, 2019
- Border-check-irrigation-design, irrigation, soil-and-water, farm-management, agriculture: agriculture.vic.gov.au, Retrieved 12 January, 2019
- Sprinkler-irrigation-design: harmonyfarm.com, Retrieved 22 March, 2019
- Easy-design-of-a-complex-drip-irrigation-system: agrivi.com, Retrieved 2 August, 2019

Permissions

We would like to thank the editorial team for lending their expertise to make the book truly unique. They have played a crucial role in the development of this book. Without their invaluable contributions this book wouldn't have been possible. They have made vital efforts to compile up to date information on the varied aspects of this subject to make this book a valuable addition to the collection of many professionals and students.

This book was conceptualized with the vision of imparting up-to-date and integrated information in this field. To ensure the same, a matchless editorial board was set up. Every individual on the board went through rigorous rounds of assessment to prove their worth. After which they invested a large part of their time researching and compiling the most relevant data for our readers.

The editorial board has been involved in producing this book since its inception. They have spent rigorous hours researching and exploring the diverse topics which have resulted in the successful publishing of this book. They have passed on their knowledge of decades through this book. To expedite this challenging task, the publisher supported the team at every step. A small team of assistant editors was also appointed to further simplify the editing procedure and attain best results for the readers.

Apart from the editorial board, the designing team has also invested a significant amount of their time in understanding the subject and creating the most relevant covers. They scrutinized every image to scout for the most suitable representation of the subject and create an appropriate cover for the book.

The publishing team has been an ardent support to the editorial, designing and production team. Their endless efforts to recruit the best for this project, has resulted in the accomplishment of this book. They are a veteran in the field of academics and their pool of knowledge is as vast as their experience in printing. Their expertise and guidance has proved useful at every step. Their uncompromising quality standards have made this book an exceptional effort. Their encouragement from time to time has been an inspiration for everyone.

The publisher and the editorial board hope that this book will prove to be a valuable piece of knowledge for students, practitioners and scholars across the globe.

Index

A

Application Efficiency, 13, 32, 102, 104, 220-223, 247
Application Rate, 50, 84-85, 102, 104
Automatic Plant Irrigation System, 61, 93-94

B

Basin Irrigation, 13, 124
Berm, 38, 40-43, 45-47, 142, 153
Border Irrigation, 123, 128

C

Canal Bed Slope, 157, 163-164
Canal Lining, 7, 25, 144, 146-147, 149-150, 153, 155, 164-166
Commanded Area, 150, 163
Compaction, 43, 46, 121, 147
Contour Bench Levelling, 138
Conveyance Efficiency, 31-32, 77, 209, 220, 222
Conveyance Loss, 53-54, 209
Crop Coefficient, 221, 223-224, 244-245
Crop Evapotranspiration, 54, 81-82, 220-221, 243-245
Crop Water Requirement, 50-51, 53, 221, 243-244
Cut/fill Ratio, 128-130, 133-134

D

Deep Percolation, 32, 35, 56, 221, 246
Deficit Irrigation, 81
Distribution Control Structures, 8-9
Distribution Uniformity, 34-35, 240, 247, 249
Drip Emitter, 96, 98, 190, 224
Drip Irrigation, 12-13, 21, 83-84, 96, 98-99, 224, 228, 241, 243, 245-246, 259-263

E

Emission Uniformity, 34, 246
End Area Method, 129, 141
Erosion Control, 8, 44

F

Fertigation, 89, 263
Field Capacity, 3, 56, 248
Furrow Irrigation, 12, 125

I

Infiltration Rate, 233-235, 239, 247-250

Intake Structure, 1, 4-5
Irrigation Canal, 66-67, 70, 144, 162
Irrigation Criteria, 55
Irrigation Cycle, 88, 245
Irrigation Efficiency, 30-31, 33, 56, 77, 86, 247
Irrigation Hydraulics, 177, 184-185, 232
Irrigation Management, 1, 55, 57
Irrigation Reservoirs, 36
Irrigation Scheduling, 2, 55-57, 60-61, 81
Irrigation System Design, 199, 224, 239-240
Irrigation Timers, 85, 87-88, 96, 99
Irrigation Uniformity, 34, 138

L

Land Evaluation, 112-113, 115
Land Grading, 112, 120, 137, 141
Level Basin, 124

M

Master Valve Control, 88
Maximum Evapo-transpiration, 54-55
Microirrigation, 33, 230
Moisture Distribution Pattern, 102-103

N

Nett Irrigation Requirement, 220-221

O

Open Canals, 6, 10, 66
Open Control Loop System, 85

P

Percolation Loss, 53-54
Permanent Wilting Point, 220
Prismoidal Formula, 129, 141

R

Reservoir Storage Efficiency, 31
River Gauging, 116-117
Root Zone, 1, 13, 30, 32-33, 35-36, 56, 81-84, 95, 121, 123, 220, 237, 239, 259

S

Seepage Loss, 74-75, 146, 150
Shepparton Irrigation Region, 248

Soil Condition, 93
Soil Maps, 39, 114-115
Soil Moisture Deficit, 245, 247-248, 250
Soil Moisture Monitoring, 3, 82
Soil Survey, 117
Sprinkler Irrigation, 33, 82-84, 102, 200, 203, 243, 251
Sprinkler Spacing, 102-103
Static Water Pressure, 251, 254
Subsurface Drainage, 39, 73, 155
Surface Irrigation, 13, 21-22, 32, 35, 48, 63, 65, 123, 125, 130, 136, 177, 222, 232-234

T
Tail Water Ratio, 35

U
Underground Pipeline System, 61, 107, 109
Uniformity Coefficients, 103

W
Water Balance Calculation, 55
Water Distribution Efficiency, 33, 221
Water Holding Capacity, 239, 245
Water Logging, 27, 150, 164
Water Table, 119, 150, 213-214
Weirs, 11, 64, 73, 172, 197-200